Fur, Fleas, and Flukes

Also by Michael Stock

The Flying Zoo: Birds, Parasites, and the World They Share

FUR, FLEAS, and FLUKES

The Fascinating World of Parasites

MICHAEL STOCK

Aevo UTP
An imprint of University of Toronto Press
Toronto Buffalo London
utorontopress.com

© University of Toronto Press 2024

All rights reserved. No part of this publication may be reproduced, stored in or introduced into a retrieval system, or transmitted in any form or by any means (electronic, mechanical, photocopying, recording, or otherwise) without the prior written permission of both the copyright owner and the above publisher of this book.

Library and Archives Canada Cataloguing in Publication

Title: Fur, fleas, and flukes : the fascinating world of parasites / Michael Stock.
Names: Stock, Michael (Terrance Michael), author.
Description: Includes bibliographical references and index.
Identifiers: Canadiana (print) 20240417682 | Canadiana (ebook) 20240417712 | ISBN 9781487509224 (cloth) | ISBN 9781487539931 (PDF) | ISBN 9781487539948 (EPUB)
Subjects: LCSH: Mammals – Parasites.
Classification: LCC QL757 .S76 2024 | DDC 591.7/857 – dc23

ISBN 978-1-4875-0922-4 (cloth) ISBN 978-1-4875-3994-8 (EPUB)
　　　　　　　　　　　　　　　 ISBN 978-1-4875-3993-1 (PDF)

Printed in the USA

Cover design: Will Brown
Cover images: Wikimedia Commons/rawpixel; Wikimedia Commons/National Library of Wales; iStock.com/tonaquatic

We wish to acknowledge the land on which the University of Toronto Press operates. This land is the traditional territory of the Wendat, the Anishnaabeg, the Haudenosaunee, the Métis, and the Mississaugas of the Credit First Nation.

University of Toronto Press acknowledges the financial support of the Government of Canada, the Canada Council for the Arts, and the Ontario Arts Council, an agency of the Government of Ontario, for its publishing activities.

Canada Council for the Arts   Conseil des Arts du Canada

ONTARIO ARTS COUNCIL
CONSEIL DES ARTS DE L'ONTARIO
an Ontario government agency
un organisme du gouvernement de l'Ontario

Funded by the Government of Canada   Financé par le gouvernement du Canada

This book is dedicated to members of the Canadian Wildlife Service, to provincial Fish and Wildlife divisions, and to national and provincial park wardens, who are devoted to protect and preserve our natural resources, particularly wild mammals, birds, reptiles, amphibians, and fish. These biologists work tirelessly, often in very trying conditions, to conserve the wonderful natural heritage that belongs to us all.

Contents

Preface and Acknowledgments   ix

Introduction: Wolves and Worms   1

1 Pinworms, Primates, and Porcupines: How Parasites Traveled the World   9

2 Stone Cold Killers: *Trichinella* in the Arctic   21

3 Who's Your Daddy? Lice on Great Apes   31

4 Giants Crawl among Us: Giant Liver Flukes   45

5 Beetles and Beavers   57

6 Stranded Whales: A Fluke Accident?   71

7 How the Zebra Got Its Stripes   81

8 Ornaments and Parasites   95

9 The Night of the Vampire: Parasitic Mammals and Bat Bugs   113

10 Your Brain on Worms: Nature's Biological Weapon   127

11 The Tale of the Tape: The World's Longest Parasite   139

12 Death by Raccoon   149

13 Moths, Sloths, Tears, and Blood   161

14 The Manchurian Parasite   175

15 A Ghost of a Chance   193

16 Sex and the Single Schistosome   205

17 The Trickster: Coyotes and Their Parasites   221

18 Fleas: The Inside Story   239

Conclusion: The Greatest Show on Earth   255

Notes   267

References and Additional Readings   271

Index   297

Preface and Acknowledgments

> I cannot persuade myself that a beneficent and omnipotent God would have designedly created parasitic wasps with the express intention of their feeding within the living bodies of Caterpillars.
> – Charles Darwin, *Life and Letters of Charles Darwin*, vol. 1

I never set out to be a biologist, let alone a parasitologist. I wanted a scientific career where I would have a chance to work outdoors, and geology really appealed to me. But I recall a student in my first undergraduate biology course asking the professor a question, and his answer stunned me: "That's a good question, and right now, we just don't know." In my other introductory science courses (chemistry, geology, physics), invariably the professor would confidently give definitive answers or tell students how they could deduce solutions themselves. Biology was different – it was a newer science, it was complex, and there seemed to be many more unanswered questions than explanations. I was hooked.

Concentrating on ecology (so I could do field work), I took a curious course that dealt with symbiotic interactions between organisms, where one partner benefited at the expense of another – parasitology. The instructor was enthusiastic and captivating, and although the topic was often gruesome (after all, it dealt with things like tapeworms and blood flukes), it was fascinating. I discovered that, in this specialization, I could study any facet of the great world of zoology. I could work with both vertebrate animal hosts (I loved birds and mammals) and invertebrate parasites (I loved insects and flatworms). I could work in any habitat – the Arctic, tropical rainforests, or even marine ecosystems. I could approach questions from the smallest cellular scale (immunology and pathology) to the largest (population biology and community ecology). I could do research using an experimental scientific method in the lab, or a field-based discovery method, or even computer modeling. Even better, parasitology had practical applications (diagnosing and treating infections or helping with wildlife conservation, for instance) but also could be used to explore theoretical ideas, especially in ecology and evolutionary biology.

Over time, I had the wonderful opportunity to do research on lots of pure and applied projects, such as how crowding might affect the daily migration of rat tapeworms; how chewing lice are distributed on ptarmigan; what kinds and abundances of parasites occur in deer and other cervids, and whether there are any biogeographic patterns; the community ecology of parasites in grebes;

whether parasites compete with each other in hosts; how parasites in hissing cockroaches may cause type II diabetes; and the health status and zoonotic[*] potential of urban coyotes – just to name a few.

My study of parasitology, which eventually landed me a teaching job, led to world travels and adventures. I got to band geese in the Arctic, walk on trails in the rainforests of Australia and the Amazon, capture and radio-collar pronghorn antelope kids in southern Alberta, necropsy an orca whale at Bamfield Marine Station on the west coast of Vancouver Island, collect rattlesnakes in Arizona, trap bats in mist nets in a cloud forest in the Ecuadorian Andes, and visit the Galapagos Islands (five times!).

Through the years, another part of this amazing discipline also emerged. The central idea of biology is evolution by natural selection, and as Darwin surmised, in no other part of biology is this idea better demonstrated than in the unbelievable adaptations and counter-adaptations between hosts and parasites. These discoveries still fascinate me today. The great biologist Edward O. Wilson claimed that there are two different ways of doing biology: either you can ask a question and then investigate it by selecting whatever organisms will lead you to an answer, or you can fall in love with an organism (in his case, ants) and then spend your life trying to find out everything you can about it. My study of parasitology let me take the second approach – I am enthralled by parasites of all kinds.

---

[*] Zoonosis: an infection transmitted from animals, especially vertebrates, to humans.

My goal in writing this book is to introduce you to a world that is going on around us that most know nothing about – wild mammals and their parasites. Today, even if you live in a major city and seldom get a chance to visit national parks or wildlife reserves, you encounter wild mammals. In my suburban backyard, I can see noisy red squirrels having territorial disputes and stealing bird seed from my feeders, jack rabbits and snowshoe hares eating my plants, and bats hawking for mosquitoes and moths around my porch at night. It's not unusual to see white-tailed deer, skunks, porcupines, and coyotes in my neighborhood, and raccoons and opossums get into my family's garbage in the city where I was raised. We all see these animals, but few of us are aware that on the inside and the outside of them live an amazing diversity of other living things – parasites. What's more, these parasites play crucial roles affecting the ecology, behavior, and evolution of their wild mammal hosts.

In this book, I hope to show you how these unseen organisms touch the lives of wild mammals. I will explain how parasites have affected the appearance, behaviors, distributions, and even the evolutionary histories of their hosts. I will point out how parasites can spill over into our pets and, disturbingly, can cause diseases and even deaths in us. I will discuss how the unprecedented environmental changes that are happening on our planet may alter the ecological balance between mammals and their parasites and how this shift could also impact us.

Selfishly, I also wrote this book to pay tribute to the remarkable people I have had the great privilege to learn from during my career. The scientific discoveries of many of these people are mentioned in the book, while others shared their expertise on mammals, genomics, and other topics. Some of these colleagues were professors and graduate student supervisors; some were students or technicians; some were dedicated professional wildlife biologists and conservationists; and some were educational colleagues that I have worked with and asked a thousand questions of. But to me, all of them were mentors and teachers. The list is too long to include here, and I'm sure I would hurt or offend someone by forgetting to include them if I did try to make a list, but please know that I could never adequately thank you enough for the education and inspiration you provided. Because of you, I was lucky enough to get to study parasites.

I also want to acknowledge my daughter Heather, who gave me technical help, proofread parts of the manuscript, and was a constant source of encouragement; Gordon Youzwhyshyn, who read a first draft of the manuscript and made suggestions for additions and deletions; the three anonymous reviewers, who critiqued the book and asked some intriguing questions; and, particularly, Stephen Jones, Carolyn Zapf, and Lohit Jagwani, at the University of Toronto Press, who worked tirelessly to improve the readability and clarity of this book. Of course, any errors or omissions are entirely my own.

Introduction: Wolves and Worms

Hidden in the dappled shadows of aspen poplar trees near the shore of a frozen lake in northern Canada, a small pack of gray wolves trot along in file on a game trail they've used to patrol their territory for weeks. Silver frost crystals on their muzzles and the steam of their breath are evidence of the clear −20°C (−4°F) day that's about to dawn. By scent, sound, and guided by two noisy ravens, the pack knows that ahead a cow moose and her calf are browsing on willow stems poking through the drifted snow.

The moose see the wolves as they break out of the forest into the open. The cow alerts her calf, and they start to flee using their long legs and large splayed hooves to glide through the snow. Although they don't seem to be moving any faster, the wolves resolutely close ground on their prey. The pack tested several different moose in the last few days and was never able to get close enough for a kill, but somehow, something is different this time. The cow trots for a short distance, then stops. Her chest is heaving, and her head is down. The yearling calf slows and tries to encourage her mother to run, but she just can't keep up.

The labored breath of the cow moose even smells different and strangely attractive to the wolves. Soon, a large female wolf has gotten close enough to the cow to lunge at her hindquarters. Within a minute the cow is down in the snow – overcome by the slashing and crushing jaws of the pack. All the yearling can do is run away.

The wolves begin their feast by eating the moose's visceral organs. They don't notice the fluid-filled brown and white capsules in the lungs of the cow, each containing a dollop of "sand." Later, as the wolves lay around idly digesting their meals and the ravens pick at leftovers, hundreds of tiny, microscopic heads that were inside each capsule (the sand) emerge from little sacs. At the end of each of these heads occur a circle of recurved hooks, and four small muscular suckers, which allow the heads to attach to the inner lining of the wolves' small intestines.

We will learn more about tapeworms in chapter 11, but suffice to say that, over the next few weeks, the worms will grow, and each will produce a number of reproductive body segments. Inside the wolves, sexually mature tapeworms reproduce, and soon each wolf that ate a capsule will be passing hundreds to thousands of worm eggs with its feces – seeding the forest floor with parasites. When the eggs, which can easily resist freezing for months, are accidentally eaten by a vole, a rabbit, a deer, or a moose, they will go on to produce golf-ball-sized, fluid-filled capsules (called hydatid cysts) in the lungs and other visceral organs of their hosts.

Over time, as the cysts enlarge, infected animals like the cow moose will not run as fast or as far as they once did; based on humans who have been infected, these animals will have sharp, burning chest and abdominal pains. Before too long, the infected hosts will become easier and easier targets for wolves and other predators, who will then go on to consume the hydatid cysts in their kills. The life cycle that ties together wolves, their prey, and the tiny tapeworms has now been completed.

An Intimate Relationship

In this system, this parasite (*Echinococcus canadensis*)[*] interacts with different mammals in two different ways. In wolves, the tiny adult tapeworms are harmless. The amounts of nutrients and calories of energy that the parasites steal from wolves are so insignificant that the little tapeworms are mostly ignored. Wolves probably mount a defensive response to the infection (a branch of the immune system that is housed in the intestine is likely stimulated), but the worms have evolved ways of dealing with this problem so that they can live long enough to mature and reproduce. Curiously, it is likely that the biggest, healthiest wolves will host the largest number of tapeworms because they are the most successful hunters.

---

[*] Previously identified as *E. granulosus*.

In prey like moose, the story is very different. The larval stages of the tapeworms, hydatid cysts, cause enormous damage. The cysts are quarter- to golf-ball-sized and take up space reserved for normal, intact organs. As a result, the physiology, stamina, behavior, and health of prey are affected, making tapeworm-infected hosts more likely to become wolf food.

Ecology is a branch of biology that tries to explain biodiversity – what kinds of organisms live in particular places, and how many there are. Biodiversity is affected by abiotic, physical features (climate, humidity, temperature) and by biotic factors – other organisms that are present. It gets complicated, however, because these abiotic and biotic factors interact. For instance, when predators (wolves) were reintroduced to Yellowstone National Park in Wyoming, from where they had been previously eliminated, people soon noticed changes to the physical landscape.

Plant life changed in the park because wolves had decreased the populations of browsers and grazers like deer, elk, and moose. This decrease led to changes in insect biodiversity because the changes to plants provided different pollination and food opportunities, which in turn led to changes in bird biodiversity and to differences in the abundance and distribution of mammals such as rodents and rabbits.

Like a pebble dropped into a pond, one biotic change in the park ecosystem (reintroducing wolf predators) caused ripples of changes that even affected the very nature of the landscape itself. The "keystone" change was the reintroduction of a predator. Predators hunt and kill prey. They get their energy and carbon (for growth and reproduction) in one short, violent act – examples like lions killing zebras and sharks feeding on seals are well known to most of us. However, far more important and far more numerous are other creatures that are also getting their energy and carbon from other living things but in a much less obvious way – parasites.

Parasites versus Predators

Unlike predators, which are usually large and have only brief, flashy, violent encounters with their prey, parasites are hidden. They tend to be small (much smaller than their hosts) and have much longer and more intimate associations with other animals than do predators. Sometimes parasites kill their hosts, but usually they quietly steal nutrients, day after day, eventually causing illness, disease, and poor condition. The main evolutionary objective of parasites is to live inside or on the outside of their hosts long enough to reproduce and then to solve the problem of successfully getting their offspring into (or onto) another host. To reach these goals, parasites often affect the normal behaviors, social lives, reproduction, ecology, and even the evolution of their hosts.

Sometimes distinguishing between parasites and predators is hard. For instance, mosquitoes are much smaller than their hosts, and females absolutely need a blood meal before they can lay eggs, but mosquitoes have only brief encounters with other animals and often feed on many different kinds of hosts/prey. Like classic parasites, mosquitoes don't kill their hosts outright, but they can spread diseases like malaria and yellow fever, which certainly do kill. Are mosquitoes and other biting flies parasites, or are they small (micro) predators? In this book, we will see examples of organisms (like *Echinococcus* tapeworms) that are slam-dunk parasites, but we'll also see organisms that are on the boundary between being parasites and predators (like vampire moths and bats).

Unlike the human mind, which has evolved to look for patterns and to organize things into categories, nature doesn't. One kind of interaction like predation can blend into another, like parasitism, and sometimes can even be considered to be mutualism – the interaction where two organisms benefit each other. For instance, *Echinococcus* tapeworms harm prey animals and are parasites, but they also help wolves get prey more easily and so can be considered mutualists.* Strangely, this phenomenon means that *Echinococcus* tapeworms may be partly responsible for changing the Yellowstone ecosystem too.

Why Parasites Are Important

The natural drama that played out in the wild at the northern lake, involving three actors – wolves, moose, and worms – is not rare or uncommon. Parasites are found in

---

* Mutualism is the interaction between two or more different organisms in which the fitness of all partners increases. The most famous example is pollination, where flowering plants get fertilized by animals that gain nutrients.

all ecosystems. Parasites have evolved many kinds of life cycles and amazing strategies in order to give themselves the best chances to pass copies of their genes to the next generation. Natural selection has produced hundreds of thousands of different kinds of parasites that exploit all kinds of hosts, from bacteria to blue whales. Because most living things are infected by more than one kind of parasite, there are more parasite species in the world than there are hosts! Wild mammals (like all hosts) must live with their parasites and have had to evolve ways to try to control them. The story of adapting and counter-adapting goes on for as long as the actors exist – but their roles can change over time.

In this book, I will share with you stories of the different kinds of parasites that occur in wild mammals. Most of my examples will involve North American hosts, with which I am most familiar, but sometimes the important role that parasites are playing is best demonstrated by examples from around the world.

Today, due to the COVID-19 pandemic that we are living through, we can see that the world's most accomplished parasites are virus particles; however, they are not living things but rather packages of chemicals (proteins and nucleic acids) that must invade live cells in order to multiply. Many bacteria and single-celled organisms are also accomplished parasites. They cause horrid diseases like malaria, sleeping sickness, and Lyme disease. Here, my stories will involve parasites that are animals (different kinds of worms, insects, mites, ticks, and others) that infect (live inside) or infest (live on the outside of) wild mammals. I hope to show you that these mostly ignored and little-noticed creatures affect whole ecosystems. Parasites have changed the evolution of mammals and are doing so even today.

It is my intention in this book to introduce you to the diversity of wild mammal parasites and to open your eyes to a world around us that few are aware of. I will show you how wild mammals have come to live with a wide variety of parasites and how the ecology, behavior, appearances, and evolutionary histories of wild mammals are shaped by parasites. I will show you examples of the unbelievable strategies parasites use to infect their hosts. Finally, I will point out the role that humans are playing in this drama, how, by changing natural ecosystems and our climate, we are affecting the lives of other mammals and their parasites, and how these changes may affect our future too.

1

Pinworms, Primates, and Porcupines: How Parasites Traveled the World

One very interesting dimension to the study of parasites of wild mammals is that it can tell us a lot about their hosts – for instance, the health status, habitats, food habits, and other behaviors. Perhaps even more surprising, parasites can also tell us about the evolutionary past, origins, and even travel history and method of transportation of host mammals. For example, roundworms (also known as nematodes) are usually small animals that are often common in "interstitial" habitats – the spaces between sand grains and sediments. The adaptations nematodes developed to live there were ideal for them to evolve to become parasites. This chapter is about some small roundworms occurring in primates, squirrels, and porcupines that tell amazing stories about the histories and world travels of their hosts.

Things That Go Bump in the Night

Deep in the recesses of the last part of the digestive systems of most mammals (but not dogs and cats) live small roundworms (phylum Nematoda), known scientifically as oxyurids. Female worms have slender pointed tails, which provide the basis for their common name – "pinworms" (Fig. 1). One type, *Enterobius vermicularis*, infects humans. The infection normally causes no significant harmful effects, but occasionally for parents of small children, the infection can be a nightmare – pun intended – because children with heavy infections of pinworms (hundreds of worms can be found in a child, and each female worm may deposit up to 16,000 eggs *per day*) experience restless sleep and have bad dreams. These infections can lead to irritability, loss of appetite, and tantrums (both for children and their parents). Symptoms are the result of female worms crawling out of the child's anus at night and depositing eggs in the perianal folds, after which the worms crawl back inside. I still remember, as a child, my mom making me take "worm medicine" – horrible, chalky-tasting stuff. I wondered at the time why I had to take medicine meant for earthworms.

Pinworms are unusual in the world of parasites because they have a genetic system of sex determination called haplodiploidy. In most animals, including humans, there are two sets of chromosomes in each cell (diploid), and two of these chromosomes are called "sex chromosomes" – the X and Y chromosomes. In mammals, having one X chromosome and a Y chromosome results in males, while having

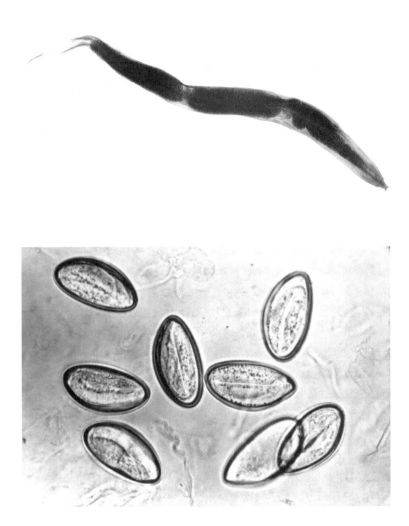

Figure 1. The human pinworm, *Enterobius vermicularis* (adult female above and eggs below). Sources: top: Carolina K. Smith MD / stock.adobe.com; bottom: courtesy of the Public Health Image Library, Centers for Disease Control and Prevention.

two X chromosomes (and no Y) results in females. Not so with pinworms.

In pinworms, females are diploid (two copies of each chromosome), and they develop from fertilized eggs. But males are haploid, having only one of each chromosome rather than two, and develop from unfertilized eggs. The development of adult animals from unfertilized eggs is called parthenogenesis – "virgin birth." In this system, where sex is determined by the number of chromosome sets instead of by special X and Y chromosomes, male worms have only a single copy of most genes, so any genes that have bad mutant alleles (forms of a gene) are quickly weeded out of the population.

This system of sex determination in pinworms has led to some remarkable population effects. For example, high levels of inbreeding are tolerable, and the pinworms within an individual host can be very closely genetically related. Also, it means that only a few individuals (in some cases just one female) can successfully colonize a host and then establish a population. Consequently, infections can spread fast.

Adult pinworms are restricted to living their lives inside a single host (they quickly dry out and die when outside a host), so infections are usually spread to other hosts living in close contact. Maybe that explains why dogs and cats are not usually infected – they tend to be solitary or live in small groups, so the transmission of pinworms from host to host is harder. Pinworms are parasites that have a "direct life cycle," meaning that infections are spread between hosts within which the parasites reproduce sexually (we'll come across other methods of transmission later). Infection occurs when small, lightweight pinworm eggs are accidentally ingested. Because of this method of transmission, most human infections with high numbers of worms occur in people concentrated together in prisons, mental institutions, daycares, and schools. An infected host can reinfect themselves with their own worms by ingesting eggs. Children often put their fingers (with pinworm eggs attached) into their mouths.

The Consequences of Parthenogenesis

These reproductive and transmission features (parthenogenesis and direct transmission) mean that pinworms have become adapted to infect only one type of host – they are parasites that show a very high degree of host specificity. Human pinworms can only infect humans and will not usually survive and reproduce in any other hosts.

This high degree of host specificity seen in pinworms led a Canadian biologist named T.W. Cameron in the 1920s to propose that each species of host should have its own species of pinworm and that pinworms could possibly testify about the evolution of their hosts. He suggested there should be a close correspondence between the evolutionary histories of mammals and their parasites. This phenomenon of congruent patterns of evolution is today referred to as co-evolution.[*] If the idea were true, then comparing pinworms

---

[*] Coevolution (the evolution of one organism affects the evolution of another and vice versa) is comprised of coadaptation (inherited features in both organisms due to coevolution) and cospeciation (as one organism diverges into two, so too does another organism).

of different hosts could give us insights into the evolutionary past, pointing out hosts that are close relatives and, perhaps, when and where different host species originated.

In 1929, Cameron claimed that, if you looked at the types of pinworms found in different primates, one species of worm would be restricted to a single genus of host rather than showing a strict one parasite species to one species of host relationship. For instance, Cameron thought that modern humans (*Homo sapiens*) and close relatives (*Homo habilis* and *Homo neanderthalensis*) would host the same species of pinworm, while chimpanzees (in the genus *Pan*) would have a different type of pinworm. He arrived at this theory because he thought that pinworms' rate of evolution would be slower than the evolution of their primate hosts. Intuitively, this idea makes sense because, as a new species of host evolve from their infected ancestors, each with its own unique genes and characteristics and distribution and ecology, the parasites trapped inside would have to adapt (or go extinct), which would take some time. Thus, pinworm evolution would lag behind host evolution.

Were Cameron's claims correct? Do host and parasite evolutionary histories parallel each other? If you deduce the family tree of pinworms, does it mirror that of the hosts? Do single pinworm species infect each host genus? Well, it took about 80 years to examine Cameron's claims. Over this time, researchers collected worm specimens from most species of primates, examined features of pinworms that indicated their evolutionary relationships, and used a method of analysis called cladistics to look at the most likely patterns of evolutionary histories for both worms and hosts.

The beauty of this work, in addition to giving us some answers about Cameron's claims, is that it pointed to some unexpected insights into primate evolution too.

Fellow Travelers and Hitchhikers

Primate biologists have always wondered about the evolutionary relationships between old world (Africa, Asia, and Europe) primates, like baboons, gorillas, chimps, and humans, and new world (Central and South America) primates, like tamarins, marmosets, and spider monkeys. When did these two groups of primates get separated? Who was the ancestor (or a relative of the ancestor)? Also, how do lemurs (found only on the island of Madagascar), lorises (found in India and Southeast Asia), and bushbabies or galagos (found in mainland Africa) fit in?

Most of the pinworms occurring in primates share features that evolved recently (called derived characters) and came from a common ancestor. These parasites are fellow travelers that formed species in concert with the evolution of their hosts. However, these "primate pinworms" also include three species found in squirrels, which are rodents. One type of pinworm is found in African ground squirrels (*Xerus inauris*); a second type occurs in Eurasian and North American tree squirrels (like the common eastern gray squirrel), which belong to the genus *Sciurus*; and a third lives in flying squirrels (*Glaucomys*) in North America. How did primate pinworms end up in squirrels? Does it indicate that primates and squirrels are recently related, or did pinworms in some primates somehow manage to infect squirrels? Or the other way around: did squirrel pinworms get into primates? Worms that do not have a history of coevolution but somehow get into unrelated hosts are hitchhikers. These hitchhiker parasites would give evidence of shared ecological habitats of different kinds of hosts but not of cospeciation. An unexpected "jump" of a parasite from one host to another unrelated host is called "host switching."

Using cladistic analysis, evolutionary biologist Jean-Pierre Hugot showed that pinworms found in the three kinds of squirrels are not closely related to each other – they got into squirrels due to at least two (and probably three) host switching events. First, primate pinworms in Africa switched into African ground squirrels; much later, some pinworms from new world primates switched into tree squirrels. This switch would mean that tree squirrels migrated from North America into the Neotropics, where they shared their arboreal habitat in the trees with primates. Later in the north, some pinworms from tree squirrels would have switched into flying squirrels. This switch must have occurred in the later part of the Tertiary (about two million years ago). Interestingly, red tree squirrels, *Tamiasciurus hudsonicus*, which are so familiar to us who live in western and northern North America, seemed to have missed the boat – they are not infected by hitchhiking pinworms at all (or, less likely, their pinworms have gone extinct).

But what about the pinworms that are specialists in primates? According to Cameron, primates and pinworms should show correspondence that reflects their evolution or association by descent. This claim has been well supported. Pinworms occur in three genus groups (*Lemuricola, Enterobius,* and *Trypanoxyuris*), which perfectly relate to lemurs and bushbabies, old world primates (including great apes), and new world monkeys. However, Cameron suggested that cospeciation would result in each primate genus having its own pinworm species (again, because worm evolution would lag that of their hosts). This idea turned out not to be supported – each species of primate seems to have its own species of pinworm. Therefore, cospeciation is even tighter than what Cameron predicted, and the evolution of new species of pinworms is "fast" – likely because of the parthenogenetic reproduction and direct transmission that pinworms have.

What This Story Tells Us about Our Evolution

Biogeographically, pinworms indicate that slow lorises (or their ancestor), which now occur in Southeast Asia, and the ancestor of lemurs, which live on Madagascar, lived in the same area. There was a parasite host switch from lorises to lemurs that probably occurred about 62 million years ago (mya). Lemurs reached Madagascar from Africa (probably by rafting on mats of vegetation) about this time and then diversified and speciated on Madagascar, as did their pinworms, about 54 mya.

All evidence of pinworm-primate-squirrel evolution indicates that pinworms originated in the African part of the ancient supercontinent of Gondwanaland in primates. Pinworms then started their journeys around the globe inside their primate hosts near the start of the Tertiary, about 65 mya. The first split (divergence) was likely between the parasites of primates and African ground squirrels (however, this idea has a problem – no squirrel fossils are known from Africa before the Pleistocene, 18 to 20 mya). Later, pinworms of lemurs and lorises diverged from old world primates. Miraculously, new world monkey pinworms show that their hosts came from African primates long after South America and Africa split apart, the most likely explanation being that pinworm-infected African hosts floated across the Atlantic Ocean on rafts of vegetation. Once isolated in South America, new world monkey pinworms diversified along with their hosts. Later, some of these worms got into squirrels that were visiting from the north. This host switch was helped because monkeys and squirrels shared habitats in trees. But due to the strong host specificity that pinworms have, contact between the two kinds of hosts must have been frequent or must have gone on for a long time. Based on what we know from human pinworm infections, resistance to infection in adult mammals can be overcome if there is repeated exposure to lots of worm eggs – sometimes preschool and kindergarten teachers get transient infections. Human pinworms, *Enterobius vermicularis*, are most closely related to those of chimps and formed as our own parasite after we diverged from a common ancestor in Africa. Now, due to world travels and our wandering ways, we have carried these pinworms to every corner of the globe.

Porcupine Pinworms Tell a Similar Story

The thought of some pinworm-infected African primates rafting across the expanse of the Atlantic Ocean to South America over weeks or months may seem improbable, but another pinworm story seems to confirm that this scenario is very possible. Porcupines are rodents that browse in trees but reside in dens on the ground. Old world porcupines (family Hystricidae) live in Africa, Europe, and Asia, including Southeast Asia, China, and India. New world porcupines (family Erethizontidae) occur from the Arctic to Mexico, Central America, and South America. The fossil evidence indicates that porcupines originated in the old world, but their pinworms tell us a lot more.

Just like primates and their pinworms, the pinworms of porcupines show a pattern of coevolution. All porcupine pinworms belong to just one genus (*Wellcomia*), and when the evolutionary history of these rodents is matched to the history of pinworms, as Jean-Paul Hugot pointed out in 2002, they are almost a mirror image.

Pinworm cladistics tells us that old and new world porcupines are closely related and also related to South American rodents called pacaranas. The worms confirm that porcupines arose in the old world, but most exciting is the evidence that suggests South American porcupines originated from African porcupines and also implies they rafted on vegetation across the Atlantic at least four times during the Tertiary, about 60 mya! Therefore, pinworms in two different kinds of hosts, primates and porcupines, tell of long distance journeys across vast oceans on mats of vegetation.

Pinworms as Souvenirs and Postcards

Human pinworms have been causing nightmares in children for a long time. The characteristic eggs (shaped like rounded footballs that are going flat on one side; Fig. 1) were found in coprolites (fossil poop) at Danger Cave, Utah, dating to 7800 BC. The physician Hippocrates wrote about them in 430 BC. Today, pinworms are perhaps the most common parasites of people – some estimates suggest that more than 200 million people are infected, over 30 percent of whom are children aged 5 to 10 years.[1] The prevalence of pinworm infection (percentage of people infected) is as high as 61 percent in India, and about 40 million in the United States are infected.[2] Some researchers have suggested that human pinworms are currently forming (or have already formed) another new species (*Enterobius gregorii*), which is adapted to infecting us by aerosol transmission (inhaling the eggs) rather than by direct ingestion. This change may have happened from times when we lived in social groups inside caves. Thus, the evolutionary history of pinworms tells many stories. Pinworms testify about where we came from and where we have been, about great ocean voyages, and about how primates, squirrels, and porcupines have evolved over the ages. These parasites are living souvenirs and postcards that document world travels.

2
Stone Cold Killers: *Trichinella* in the Arctic

Today we are living in a world that is undergoing global climate change, including rapid warming of the polar regions. In the Arctic, permafrost is thawing, the extent of permanent sea ice is shrinking, and huge parts of the glacial continental shelves of Greenland and Antarctica are breaking off. How all kinds of living things respond to these changes in the Arctic is unknown. Some may extend their ranges into the north from more temperate habitats; some Arctic residents may have their populations reduced; and some may go extinct. However, there is one parasite that has, over its evolution, demonstrated a remarkable ability to adapt and to exploit ecosystems ranging from the tropics to the poles. No one knows yet how this parasite will respond to the pressures of climate change. But because it uses an unusual transmission strategy, I think that this parasite will easily survive and, perhaps, thrive. The strategy this worm employs, which has made it so successful, is to painfully kill its hosts.

Trichinosis: A Nasty Encounter with a Deadly Parasite

In October 1943, during the height of the Second World War, a secret weather station, the Schatzgräber, was established by the Nazis in Franz Josef Land, an archipelago in the Arctic Ocean located around 81° N latitude off the coast of Arctic Russia. The unsupplied station was occupied by a party of eight to ten men. In late spring of 1944, they killed a polar bear (*Ursus maritimus*) for food, and all but one of the men (who was a vegetarian) consumed some of the meat rare in an Arctic version of steak tartare. Two or three days after their polar bear meal, the men developed diarrhea, nausea, abdominal pain, and fatigue. About a week later, they started to experience fever, facial swelling, weakness, headache, pink eye, and muscle pain with tenderness, and swallowing and chewing became painful. Their leader, a man named Lieutenant Markus, contacted the German navy, and a plane was sent to their rescue. The evacuation mission was very dangerous, and the wheels of the plane were damaged on the icy island when landing and had to be repaired before takeoff. With a doctor and all the Schatzgräber men on board, the plane was able to land in Nazi-occupied Oslo, Norway, where members of the team were hospitalized for several months. Subsequently no one died. The cause for all of the Germans' misery, and for abandoning the weather station, was a small parasite they got from eating bear meat, *Trichinella nativa*, and the disease was trichinosis.

A Parasite Evolved to Kill

Unbeknownst to the men of the weather station, when they ate the polar bear meat, they also consumed the encysted juveniles of small roundworms (nematodes) that belong to the genus *Trichinella* (Fig. 2). After the men ate the cysts, their digestive juices released the juvenile worms from capsules; a day later, in the small intestine, the worms matured to adult males and females. The small adult worms (2–3 mm or 0.08–0.12" long) burrowed into the lining of the men's intestines. After mating, each female worm deposited 500 to 1500 juveniles in the intestinal lining. The juvenile worms have small sharp stylets that they use to penetrate the intestine, which was the cause of the intestinal symptoms the men suffered. The species of *Trichinella* the men were infected by (*T. nativa*) are known to cause prolonged digestive problems and diarrhea. Once through the gut lining, the worms entered the men's lymph vessels and circulatory systems, and migrated throughout their bodies.

When the migrating juveniles reach capillary beds, they burrow through the vessels into striated skeletal muscles. Although they may enter any muscle cells, they seem to prefer muscles in the upper thorax, jaw, and neck regions, including the diaphragm and intercostals (used for breathing) and masseter, digastric, and tongue muscles (used for chewing and swallowing). The juvenile parasites are amazing

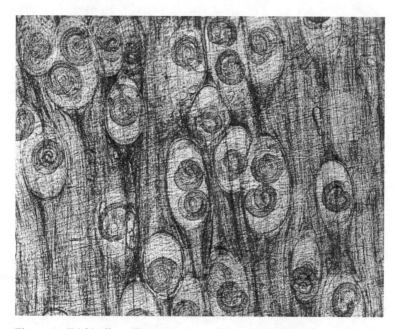

Figure 2. *Trichinella nativa* cysts – juveniles in host muscle nurse cells. Source: Image Source Limited / Alamy Stock Photo.

because they cause muscle cells to effectively become zombies. To make this transformation, *Trichinella* worms have cells called stichocytes, which produce proteins that invade host muscle cells. The parasites reprogram muscle cells for their own survival.

First, the muscle cell nuclei enlarge, and the amount of smooth endoplasmic reticulum membranes inside the cells expand (the parasites are prepping the cells to make and secrete lots of proteins). Next, muscle cells lose their contractile proteins (actin and myosin), and mitochondria are destroyed and replaced by smaller versions – the muscle cells can no longer do their normal job. The parasites then make the host cells secrete lots of collagen protein, which forms protective capsules around the cells that contain worms. Juveniles seize control of the host cell's genes and force them to increase production of a protein called vascular endothelial growth factor (VGEF), which stimulates little blood vessels to grow into the collagen cysts. These new vessels provide nutrients to feed the parasites. Finally, host cells are held in arrested development as their normal cell division cycle is stopped. These altered zombie host muscle cells can no longer contract but act as nurseries for the parasites – they are in fact called "nurse cells."

Inside nurse cells, parasite juveniles eat and grow so that, in four to eight weeks, they are 1 mm (0.04") in length (the largest known parasites that live inside host cells). Now the worms are ready to infect another host. Inside nurse cells, juveniles go into a resting state of halted development and use an anaerobic metabolism, so they require no oxygen. The worm cysts that were in the polar bear muscles that the Germans ate, *Trichinella nativa*, are an arctic-adapted species that can survive in a frozen carcass for five years. In Lieutenant Markus and his men, the cysts would eventually stimulate a long-standing immune response, leading to the death and calcification (hardening into calcium salts) of the juvenile worms, but this process would take months to years and would result in lifelong muscular pain and soreness.

When juveniles were migrating in the men's circulatory system, they damaged blood vessels, resulting in the edema (swelling) the men had in their faces and hands.

Lost, wandering juveniles can cause a variety of problems, including inflammation of the lining of the lungs and meninges (linings of the central nervous system), resulting in pneumonia and meningitis, as well as kidney damage and brain and eye problems, like the conjunctivitis (pink eye) the men had.

Occasionally, juveniles enter heart muscle cells and cause myocarditis (inflammation of the heart muscle) and areas of tissue death (necrosis). The fever experienced by the men in the weather station occurred during the worm migration phase.

When the worms started to penetrate muscles, they caused intense pain and made it hard for the men to breathe and swallow. They may have had low blood pressure, heart damage, and neurological problems, including deafness, blindness, and hallucinations. Death (which the Nazis escaped) is usually due to kidney damage, pneumonia, or heart failure. Because these parasites are transmitted when muscles of an infected host are eaten by another mammal, either by carnivory or by scavenging, the massive body damage and death that *Trichinella* causes helps the parasites to get passed to another host – the pathology to one host improves parasite passage to another host.

An Arctic Mystery

The lifecycle of this arctic parasite is mysterious. Polar bears probably get infected mostly by scavenging frozen carcasses. Dead carnivores (like wolves or other bears) would be a likely source of parasites, but studies suggest that polar bears avoid eating bear carcasses (and risking infection) unless regular food, like seals, is scarce.[1] Polar bear cubs do not get infected with *Trichinella* until they are one year old, but anywhere from 50 to 80 percent of adult bears in any population are likely to be infected, and polar bears where the German weather station was located in Franz Josef Land are at the upper end of that range of infection.

Polar bears may also get infected by eating seals or walruses. Transmission of worms among terrestrial mammals, especially carnivores and scavengers, is easy to understand. But how do marine walruses and seals get infected? Walruses eat mostly bottom-dwelling invertebrates, like clams, mussels, snails, and crabs; however, fish may occasionally be consumed too. Arctic seals (like ringed seals, bearded seals, and harp seals) eat mostly fish, but some invertebrates are also consumed. If a fish or crustacean has scavenged on *Trichinella*-infected mammal flesh, these animals may serve as temporary transport hosts – the worms stay alive in them without developing. Regardless of how walruses get infected, their meat has been a common source of human infection for Indigenous people in the Arctic. For example, since 1975, 41 percent of 241 cases of trichinosis in Alaska were attributed to ingestion of walrus.[2] Juvenile worms in cysts are killed by cooking meat, but traditional foods using raw or fermented meat can lead to infection. Although we do not understand its arctic life cycle completely, there is no doubt that *T. nativa* is a very successful parasite of the far north.

A Versatile Survivor

Roundworms belonging to the genus *Trichinella* have an interesting evolutionary history, which has resulted in some being particularly well adapted to thrive in the Arctic. The worms likely originated in Central Asia about 281 million years ago (mya). Their closest relatives are whipworms, *Trichuris* spp., which occur in the intestines of many mammals, including humans, ruminants, and canids. From Central Asia and Eastern Europe, *Trichinella* spread throughout the entire world because it could adapt to all kinds of ecosystems – it is a versatile survivor. For instance, in Africa, carnivores and scavengers such as big cats and hyenas eat warthogs that are infected with types of *Trichinella* adapted to torrid conditions. *Trichinella* also spread to Western Europe, Southeast Asia, and Australia, where it uses small carnivores and rodents, and then it spread throughout North America into South America, as far south as Patagonia. The parasite is remarkably adaptable and has responded to climate changes and habitat changes throughout its history. By 2 mya, it diversified into strains like *T. nativa* that infect arctic predators and scavengers, and it became a fixture in that harsh ecosystem.

How Will Climate Change Affect *Trichinella*?

Global climate change is greatly affecting the entire Earth but is especially strong in arctic areas. Sea ice is rapidly disappearing, and the permafrost tundra is thawing. Many temperate organisms are extending their ranges northwards. How might *Trichinella* be affected? Will the number of cases of trichinosis in wild mammals and in humans change? If agricultural activity spreads into northern areas, where it was previously unsustainable, contact between wild and domestic animals could increase. Most cases of trichinosis in temperate areas previously were caused by *Trichinella* using pigs (which are omnivores that will scavenge on offal) and farmyard rats as hosts. There is conflicting evidence about how well arctic-adapted *Trichinella nativa* can infect pigs and rats, but potentially it may infect domesticated animals. Today, in temperate North America, most cases of trichinosis result from hunters eating undercooked bear meat, as the Germans did at the weather station.

It seems certain that, as a result of warming in the north, transmission of *Trichinella* will change. Walruses along the arctic coast of Alaska have become more pelagic (venturing more into the open sea) and feed more often on carcasses of mammals, especially ringed seals and perhaps bearded seals, than they did before. Remains of seals were found in 11 percent of walrus stomachs in 1983, up by 2 percent from the 1960s.[3] Continued warming will change communities of parasites in arctic mammals because of changes to feeding habits. For example, polar bears need sea ice to capture seals, but as ice disappears, bears may have to rely more on scavenging and perhaps cannibalism. Warming and reduced salinity of the sea will affect the abundance, types, and distributions of invertebrate animals, which may act as transport hosts that infect walruses and seals. Although we are not certain of the way *Trichinella* gets into marine mammals, it seems certain that transmission of this killer parasite will increase.

With less pack ice, the population of polar bears is sure to decline, but they may be replaced in the warmer ecosystem by black and grizzly bears, both of which are good hosts for *Trichinella*. Already, grizzly bears encounter polar bears more frequently, and there are even confirmed hybrids (sometimes called pizzly or grolar bears).[4] Feeding changes, new types of host mammals, and the altered health condition of arctic-adapted animals likely means that there will be many more opportunities for *Trichinella* to do what it has done throughout its history – adapt, survive, thrive, debilitate, and kill.

3

Who's Your Daddy? Lice on Great Apes

Lice are small, flightless insects that spend their entire lives, from eggs to adults, nestled quietly within the warm feathers or fur of their bird and mammal hosts. Lice do not survive for long off a host, so they must always be with a host – they are "obligate" parasites. Because they cannot fly, lice can only be transferred between hosts by close, direct, physical, body-to-body contact (except for some human lice, which can also use objects like shared hats, hair brushes, and combs to spread), so most lice infestations occur between parents and offspring, between siblings and playmates, and between mates.

Most lice on mammals are "sucking" lice because they take small blood meals at the skin's surface. Consequently, besides being irritating, lice are notorious for spreading dangerous, deadly bacterial and viral diseases. Due to all of these biological features of lice, we now understand that they have had extraordinary effects on lousy hosts such as the great apes. Today, great apes (family Hominidae) include eight species of primates: three species of orangutans (*Pongo*), two species of gorillas (*Gorilla*), two species of chimps (*Pan*), and one remaining species of humans (*Homo sapiens*). Lice may be partly responsible for causing the amount, types, and distribution of fur on great ape bodies, so our appearance and what we find attractive in others has been shaped by lice. Lice also may have led to many of the social interactions and other behaviors of great apes and thus for the development of our magnificent, complex brains. In addition, lice illustrate where great apes came from and how we are related to other hominids.

An Example of Why Lice Are Important: Lice and the Black Death

The year 1347 was devastating for people in Western Europe. Plague (or the Black Death) was transmitted from the northeastern part of the Black Sea to southern coastal Mediterranean trade cities such as Genoa and Marseille. Plague bacteria, *Yersinia pestis*, are thought to have originated in rodents in Asia and then spread from ship rats by fleas (*Xenopsylla cheopis*) to humans. After appearing on the European Mediterranean coast, the infection spread inland and northward; by 1350, it had killed nearly 12 million – about half the people in Europe at the time.

When fleas feed on the blood of infected black rats (*Rattus rattus*), *Yersinia* bacteria reproduce in the first part of flea intestines. After feeding, rat fleas lay their eggs in debris at the bottom of rat nests. Flea larvae hatch from these eggs, feed on organic debris, pupate, and produce the next generation of adults (see chapter 18). Because they spend so much time not living on a host, environmental conditions in the host nest or den are critical for flea development, and rat fleas need warm or moderate climates to prosper – they are barely infected with plague bacteria at temperatures below 10°C (50°F) or above 27°C (81°F). Nevertheless, when infected by plague bacteria, both rats and humans usually suffer a gruesome and painful death. Plague bacteria evolved a unique type of toxin, the ability to form colonies on surfaces (biofilms), and the loss of some genes used to colonize the digestive system only, so they were transmitted readily and could spread throughout the body. After a flea bite, the bacteria would enter lymph nodes (especially in the groin) and cause swollen, painful areas called buboes. Once the germs spread in the bloodstream, they invaded the lungs, liver, spleen, and skin and caused hemorrhages that resulted in a darkened appearance (Black Death), a high fever, coughing, and a drop in blood pressure that could cause coma and death – all in just a few days.

The rapid spread of Black Death throughout Europe has raised some interesting questions. How was the plague so easily spread from warm coastal areas into colder Central and Northern Europe, while rat fleas thrive and get infected most readily by *Yersinia pestis* only at warmer temperatures? Why did the deaths of rats go unnoticed, and how did the disease persist when the population of rats declined? Could the disease have been spread some other way? Might there have been another agent for the spread of plague? While there are no proven answers, we have a few clues as to the likely culprits.

Humans are infested by our own fleas, *Pulex irritans*, and these can transmit plague bacteria but not nearly as effectively as do rat fleas. However, there are other human ectoparasites (parasites of the skin and external surfaces) that can transmit plague better, and these parasites might have been responsible for the epidemic. Sucking lice (Anoplura) are common blood-feeding insects of mammals, including humans. We host three different kinds of sucking lice (Fig. 3): head lice (*Pediculus humanus capitis*); body lice (*P.h. corporis*); and pubic lice, commonly called "crabs" (*Phthirus pubis*). Compared to our closest living relatives, chimpanzees and gorillas, as well as to all other great apes, we are the lousiest. While these apes are infested at the most by only one species of louse each, we have three kinds of lice from two genera (*Pediculus* and *Phthirus*). That, in turn, raises its own questions: Why are we so troubled by lice? Where did they come from?

Head Lice

Head lice have been infesting us for a long time. The earliest known human case was from a Brazilian mummy, 10,000 years ago. Female head lice cement their eggs (called nits) onto hairs. After hatching from nits and growing, adult head lice live in the fine hairs of the head and cannot infest eyelashes, eyebrows, coarse pubic hair, or even auxiliary

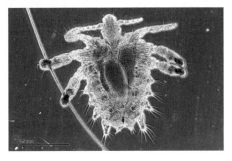

Figure 3. Sucking lice of humans: head louse (top left), body louse (top right), and pubic louse (bottom). Sources: top left: Gilles San Martin / CC BY-SA 2.0 DEED (https://creativecommons.org/licenses/by-sa/2.0/deed.en); top right: BSIP SA / Alamy Stock Photo; bottom: Josef Reischig / CC BY-SA 3.0 DEED (https://creativecommons.org/licenses/by-sa/3.0/deed.en).

hair in our armpits. This habitat specificity is due to the size of little pincers on the ends of lice legs (tarsal claws), which are used for holding onto and moving around on hosts. The claws inhibit the removal of lice by grooming but are too small to effectively allow head lice to grasp thicker hairs than those on our scalps.

Head lice are transmitted by physical contact, and many of us are familiar with them because frequent outbreaks happen to school children. Discovery of an infested child used to mean they were sent home and banned from school until delousing with insecticidal shampoos and fine-toothed combs. The same problem faced ancient Egyptians and Romans, who used two-sided combs – coarse teeth on one side for straightening unruly hair and fine teeth on the other side for removing nits and adult lice. Today, some school boards do not suspend lousy students but send them home with a note to inform parents of the issue and to advise treatment options. Untreated cases can lead to massive infestations, where the hair becomes matted, greasy, and smells. Although irritating, head lice do not seem to transmit bacterial and viral pathogens often. It is very likely that people were already well infested with head lice in the fourteenth century but unlikely that head lice were a major factor in spreading Black Death. However, the same cannot be said for body lice.

Body Lice

Body lice, which are a little larger than head lice, do not occur on the body surface of their host except to feed. They spend most of their time hiding in clothing. Body lice bites cause red papules (lesions due to an inflammatory response), which can exude lymph. The intense itching (pruritus) that the papules cause leads to scratching, which can easily worsen and lead to dermatitis (inflammation of the skin) and then bacterial infections. Over time, the skin becomes dark and thick – a condition called "vagabond's disease."

Thomas à Becket (1118–70), the archbishop of Canterbury, wore many layers of clothing as well as a hair shirt made of bristly horse hair next to his skin. When he was murdered at the altar of Canterbury cathedral, it was discovered that his shirt was loaded with biting body lice; one report claimed that the shirt moved on its own and "boiled over with them like water in a simmering cauldron."[1] Because each body louse can bite a person up to five times a day, and because the archbishop was likely infested by thousands of lice, the torment he subjected himself to is hard to imagine. Today, body lice infestation (pediculosis) most often occurs to the poor, the homeless, and refugees who cannot wash clothing and are forced to live in crowded conditions. Certainly, body lice were common during the plague years of the 1300s.

Unlike head lice, body lice spread many infections. These infections are infamous because they have been historically important – wars have been lost due to body lice and the diseases they spread. Trench fever, caused by the bacterium *Bartonella quintana*, affected Napoleon's army and also occurred among German and Allied troops who fought in crowded, muddy trenches during the First World War. Other bacterial diseases spread by body lice include epidemic typhus (due to *Rickettsia prowazekii*), relapsing fever (*Borrelia recurrentis*), and bubonic plague. Plague bacteria infect lice after just a single blood meal; they then multiply in the lice and are excreted in louse feces. Consequently, it is very possible that the Black Death was primarily spread by body lice, not by rat fleas.

The Shared Histories of Humans and Lice

Because body lice live in clothing and head lice on the scalp, the split of *Pediculus* lice into two types must have occurred recently – after we started to wear clothes. Evidence suggests that head and body lice went through a population bottleneck (a drastic decline) when hominids diverged from other apes and we lost most of our fur to become naked apes. In fact, there is a possibility that a reason for us being the only naked great ape was because of lice. Having little hair reduces the habitat for *Pediculus* lice and thus their population size. Fewer lice would lead to less transmission of microbial infections and would improve our overall health. This advantage might mean that the nakedness of our ancestors was selected for, and this selection might have been reinforced because we would learn to prefer mates with less hair since they would be healthier and not infested by as many lice, which are transmitted during mating. As a result, nakedness would become a sexually selected character.

Clothing deteriorates easily, and few archaeological sites provide evidence of clothes, but Neanderthals (*Homo neanderthalensis*) living in cold Northern Europe before *Homo sapiens* wore animal skins to survive. Based on molecular evidence of genetic mutations, *Pediculus* lice probably colonized hominid clothing about 170,000 years ago and were then on their way to becoming body lice. Although Neanderthals wore animal skin clothes, modern humans learned how to make better-fitting clothes using bone needles and better cutting tools. Our bodies were leaner than stockier Neanderthals, and we would lose body heat faster than them, so making better clothing helped us to cope with cold in order to survive and to compete. But wearing tailored clothing probably increased the prevalence (percentage of the population infested) and intensity (average number of parasites on infested hosts) of body lice. Even before we colonized colder Europe, *Homo sapiens* in Africa may have worn clothes to shield themselves from sun exposure and abrasions, or perhaps to adorn their bodies and demonstrate social status.

Nevertheless, whatever the reason was for us to wear clothing, around 170,000 years ago our body lice were ready to use the warm, moist, protected habitat that our clothes provided.

Heirlooms and Souvenirs

Studying the genomes of *Pediculus* lice collected from around the world has provided some remarkable insights about how humans originated in Africa, then dispersed. Five genetically distinct clades (organisms descended from one common ancestor) of lice have been recognized. Combining genetic data from lice with that from archaic hominids like Neanderthals has supported an idea that three waves of dispersal occurred: about one million years ago (mya), Denisovan hominids lived in Eastern Asia, from Siberia to Southeast Asia; about 600,000 years ago, Neanderthals moved into Europe and Northern Asia; and about 70,000 years ago, modern humans invaded Eurasia and later spread around the globe.

The genes of human *Pediculus* lice indicate that we coexisted and inherited lice from both *Homo erectus* and from Denisovans in Asia, and also from Neanderthals in Europe. We picked up lice from these hosts about 170,000 years ago, and then head lice diversified to colonize clothing to become body lice. It is believed that the modern human population (and, thus, our lice) experienced a drastic reduction in size about 70,000 years ago, but the genetic diversity found in both head and body lice on people today is very large. This diversity suggests that several races of *Pediculus* lice colonized *Homo sapiens* and escaped extinction when their regular hosts, such as Neanderthals, disappeared. *Pediculus* lice testify that we must have had close body contact with other hominids – an idea also supported by analysis of our genome.

More Lice

Besides head and body lice, we are hosts to another kind of sucking louse. Pubic lice, *Phthirus*, are shorter and wider than *Pediculus* lice, so they look like tiny crabs. Their tarsal claws are larger and stouter and adapted to hold onto coarse hair, such as that in the pubic area, but crab lice can also colonize auxiliary (arm pit) hair, eyebrows, and eyelashes. Pubic lice are transferred from host to host during sexual contact. Stubby and slow-moving pubic lice are less active than head and body lice, and take longer to feed. Usually the only symptom of a crab louse infestation is itching.

Of all the living great apes, we are the only type that has pubic hair – in fact, on other apes, such as chimps, the groin area usually has less fur. If the argument about us becoming naked apes in order to reduce lice infestations is correct, why would we have pubic hair with its own type of louse? One potential answer is that perhaps we have pubic hair specifically because we are otherwise mostly naked: pubic hair then acts as a visual signal that we are sexually mature and perhaps also assists in retaining odors, as an olfactory cue. Regardless of the reason for its presence, how did we come to be infested by crab lice?

To understand how humans acquired crab lice, we need to go deeper in time and look at our closest living relatives, chimpanzees and gorillas. The two different kinds of sucking lice, *Pediculus* and *Phthirus*, diverged about 13 mya. Today, chimps are infested by their own species of head/body louse, *Pediculus schaeffi* – the only other species of *Pediculus* on great apes. *Pediculus schaeffi* can be distributed all over the fur of chimps, but due to grooming, they usually exist more commonly in areas of the body that are harder to reach, which is likely one reason why allogrooming (grooming another animal) is a common social behavior. Besides providing hygiene, allogrooming also reinforces social bonds among animals and promotes social cohesion and peaceful interactions among the members of a troop. The complex behaviors involved in allogrooming require coordinated communication, so perhaps mutual grooming was at least partly responsible for the development of our intelligence. Did lice make us smart?

The divergence time for chimps and hominids, and also for the chimp and hominid species of *Pediculus*, turns out to be the same – 5 to 6 mya. Thus, our louse, *Pediculus humanus*, is a "heirloom parasite" – one we acquired by cospeciation from our ancestors. By contrast, the time of divergence between gorillas and the common ancestor of chimps and humans was at least one million years earlier, about 7 mya. So why don't gorillas also have a heirloom *Pediculus* louse too? One likely explanation is that gorillas did, but it has gone extinct.

Pubic lice, however, have a different story to tell. Gorillas are infested by their own species of crab louse, *Phthirus gorillae*. Based on DNA evidence, the divergence time between human crab lice and gorilla crab lice was just 3 to 4 mya. This divergence was long after the 7 mya divergence time between gorillas and human/chimp ancestors, so our crab lice cannot be heirloom parasites acquired as the result of cospeciation. In fact, this evidence suggests that the best explanation for the origin of our pubic lice is that we got them from gorillas! It turns out that human pubic hair is similar in diameter to gorilla fur, so *Phthirus* lice were able to colonize humans and survive.

But then why don't chimps have pubic lice as well? Two explanations come to mind. Either chimps did have pubic lice, but they went extinct, or chimps and gorillas have never had close enough contact for crab lice to be acquired. Even if crab lice did get onto chimps, their hair may be too fine for crab lice to colonize. Today, pubic lice may be going extinct in humans too because of the cosmetic practice of Brazilian waxing – less hair means fewer lice.

Our crab lice are likely an example of a "souvenir parasite" – parasites we got due to close contact with gorillas (perhaps by hunting them). This acquisition is also a good example of a "host capture" event, where unrelated hosts acquire parasites from animals living in the same ecosystem.

If the transfer of lice from one host to another (for example, crab lice from gorillas to humans) seems hard to believe, there is another louse example. New world primates (like howler monkeys) are infested with a head/body louse named *Pediculus mjobergi*. This louse's physical features are very similar to human *Pediculus* lice. When several genetic markers of monkey lice were compared to lice from people in a remote Amazonian tribe, the evidence showed that lice had likely been transmitted from humans to monkeys – the monkeys acquired a souvenir. Humans first reached South America about 15,000 to 20,000 years ago, possibly by a Pacific sea route, and people then as now hunted monkeys (and perhaps kept some captive as pets). The close contact between different mammals (monkeys and humans) allowed for the transfer of a parasite – a living souvenir – but this time from humans to monkeys.

The relationship between sucking lice and their primate hosts, especially great apes, tells many stories. By comparing physical features and differences in genes, and by looking at current geographic distributions, ecology, and diseases, we see evidence of coevolution, world travels, and dispersal. In addition, sucking lice and the diseases they transmit may have been at least partly responsible for a major distinguishing human physical feature (our nakedness) and for some of our social behaviors and our sexual preferences. More recently, lice may have played a part in wiping out half the human population of Europe during the Black Death, but their scientific study also helps us to know more about our ancestors, how they lived, and where we came from. You can learn a lot about hosts by looking at their parasites.

4

Giants Crawl among Us: Giant Liver Flukes

Today, with more than seven billion of us on the planet, we are causing major disruptions to natural ecosystems. Habitat alterations, climate change, and translocations of animals from native habitats to places where they never naturally occurred can have profound and unexpected consequences. In one particular mammal parasite we can see all of these factors at play – giant liver flukes.

Sweetmeats

A wildlife biologist once hiked into a hunting camp in the woods of upper state Michigan. The hunters were excited to brag about their great hunt and the beautiful white-tailed deer (*Odocoileus virginianus*) buck they had shot. Celebrating their success, the men generously offered to share with the biologist the best parts of the deer, "the sweetmeats," which they were sautéing in butter. Curious about what parts of the deer the sweetmeats were, he cheerfully accepted their generous offer. In amazement, when he looked at the plate they gave him, he saw several leaf-shaped medallions that he recognized immediately – the sweetmeats were giant liver flukes, large parasitic flatworms.

The liver flukes that served as the hunters' feast were *Fascioloides magna*, a species of flatworm (phylum Platyhelminthes). Flatworms include small, free-living marine and freshwater animals that usually scavenge at the bottom of their aquatic habitat and several kinds of parasitic worms, including tapeworms (cestodes) and flukes (trematodes). As their name suggests, flatworm bodies are flattened dorso-ventrally (back to belly) so they can take advantage of living on the bottom of aquatic habitats under rocks. Most flukes (trematodes) are small, about 5 mm in length by 2 mm in width (0.2" x 0.08"), and sexually mature adult flukes are parasites of vertebrates. Most parasitize fish, but they also reside in amphibians, reptiles, birds, and mammals. The majority of flukes are specialized to live in the intestines of their hosts (using a muscular sucker around their mouth and another on their body as holdfasts), but flukes can also occur in other internal organs, including the liver.

The Liver Fluke Life Cycle: It's Complicated

Liver flukes, like most flukes, have complicated life cycles that include a snail (a gastropod mollusk) as an "intermediate" host. In evolutionary time, trematode fluke ancestors were likely free-living scavengers and only later became internal parasites of snails and other mollusks. Subsequently, they went on to infect vertebrates (probably ones that ate a lot of mollusks) as final hosts, within which they became sexually mature and were able to reproduce.

Intermediate hosts are often used by parasites as a way to bridge the gap between hosts in one ecosystem (such as a freshwater pond) and another (such as a terrestrial forest); see Fig. 4. For example, when a white-tailed deer that is infected with adult giant liver flukes defecates near water, it passes parasite eggs. Inside each egg, a tiny larva called a miracidium develops. The little miracidia are covered by a ciliated surface, with numerous hair-like locomotor structures they use for swimming. If embryonated eggs get into aerated freshwater, little caps on the eggs pop open – they hatch. Development of the larvae in eggs is affected by climate, especially temperature and moisture. Cold or hot temperatures can delay development, but eggs can usually successfully overwinter in temperate areas. Eggs that are not deposited in or near water are fated to die.

Once the eggs hatch and release miracidia, the larvae use a stored energy supply to frantically swim in search of a freshwater snail. Because they do not feed and use a limited energy supply, miracidia only have hours to complete their mission and find an appropriate snail intermediate host. To help them, each miracidium has a light-sensitive eyespot and several chemical detectors. Using these, the little liver fluke larvae move toward light and follow chemical gradients produced by mucus that is secreted by lymnaeid snails (aquatic snails from the genus *Lymnaea*).

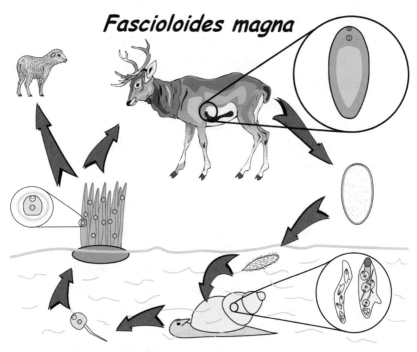

Figure 4. Life cycle diagram of a liver fluke. Source: Veterinary Parasitology Group of the College of Veterinary Medicine at North Carolina State University.

Most miracidia will die, but those that successfully find a suitable snail secrete enzymes that break down snail tissues and allow the larvae to penetrate. Inside, they go to the pulmonary sac (a moist cavity used by snails for respiration) and metamorphose to another larval stage – sausage-shaped sporocysts. All parasitic flukes have a remarkable body surface called the neodermis (*neo* – new; *dermis* – skin), a living tegument that acts as an interface between the host and parasite, absorbing nutrients like a sponge and allowing the sporocysts to grow and develop. Sporocysts have cells inside that divide and asexually produce several (up to 14) of the next larval stage called mother redia. Due to this asexual multiplication, a single miracidium inside a snail can greatly increase the number of flukes.

Mother redia are different from sporocysts because they have mouths and simple digestive systems, so they can supplement nutrients that are absorbed through their body surface by eating snail tissues. Redia go to the snail's liver (hepatopancreas) and gonads. The choice of these tissues by the parasites is not an accident – it has been selected for because castrated snails can survive without gonads (but not reproduce), and the liver is a "plastic" organ that continually regenerates as it is consumed. As a result, snails survive the fluke infection but increase their body size (they put energy usually devoted for reproduction into growth). This phenomenon has been called "gigantism." Although gigantism has not been selected for in all parasitic flukes, for giant liver flukes, it creates more space and food for parasites and keeps infected intermediate hosts alive long enough to amplify the number of parasites in the ecosystem.

After eating and growing, mother redia again clone asexually and form "daughter" redia. Like Russian stacking dolls, these larvae will eventually produce another larval stage, called cercaria – again by asexual multiplication. Cercariae will be shed from the live snail, day after day, into water. It has been estimated that one snail infected by a single miracidium will ultimately release about 1000 cercariae! This whole process, from miracidia infecting snails to cercariae being shed into water, takes about 42 to 63 days, depending on temperature.

Cercarial stages of giant liver flukes that are released into water do not feed, but they have rudimentary guts, a small muscular sucker around their mouth, a little stored energy, and a club-shaped tail. The tail is important because it lets them swim in a herky-jerky, thrashing manner until they come into contact with plant material at the water's edge. There, the cercariae drop their tails and secrete a protein coat that inhibits drying out and lets the flukes stick onto the vegetation. These larvae (called metacercariae) become metabolically inactive and play a waiting game. Surrounded by their cyst walls, the metacercariae survive in a dormant state until they accidentally get eaten by a grazing deer at the water's edge.

This complicated and hazardous giant liver fluke life cycle accomplishes two evolutionarily selected objectives: first, it lets flukes multiply their numbers, which increases the odds that a few worms will eventually get into a final host; and second, it gets flukes that are locked up inside a deer host into a habitat where another deer may feed. For all parasites, a major selection pressure that has shaped their evolution is transmission – getting offspring into a susceptible host where survival and reproduction are optimal.

Home Sweet Home

Once a grazing deer eats fluke metacercariae that are stuck onto aquatic vegetation, the cyst wall gets digested, releasing juvenile flukes in the host's small intestine. The little juveniles use enzyme-secreting glands to break down the host's intestinal wall. The worms then crawl through the intestine and enter the deer's abdominal cavity. Soon, a migration begins. Juvenile flukes use their muscular suckers and body muscles to crawl over the internal organs, drawn toward chemicals (such as glycocholic acid) in the gallbladder and bile ducts. These chemicals guide them to the deer's liver. Once there, they burrow into the liver and meander, for about 30 weeks, eating liver and growing as they go. Because there is blood in their diet, the flukes leave tracks of black pigment and spots of hematin (oxidized hemoglobin) throughout the liver. There is some speculation that single flukes continue to wander in the liver until they happen to encounter another fluke with which to pair. Flukes may use cholesterol as attractants to find each other.

Eventually, pairs of giant, sexually mature adult flukes stimulate their host to form thin-walled, fibrous capsules (called pseudocysts) around them in liver tissue. These giants are 4–10 cm (1.6–3.9") long by 2–3 cm (0.8–1.2") wide, roughly 20 times bigger than most flukes. Usually, pseudocysts contain a pair of adult flukes, sometimes more, and rarely one. Even though adult flukes are hermaphrodites, each with functional ovaries and testes, and even though flukes can self-fertilize, they prefer to cross-fertilize – presumably to increase genetic variability.

Pseudocysts in deer livers are spherical or ovoid, about 5–10 cm (2–3.9") in diameter. Inside them, besides adult flukes, are fluke eggs and a slimy, dark fluid. Hopefully, the Michigan hunters who ate giant flukes first washed off this greenish-black slime, which comes from the fluke's excretory system. Inside a pseudocyst capsule, each fluke can produce as many as 4000 eggs per day and may live for five years – a pair of worms in a capsule could produce more than 14 million eggs in their lifetime. Embryonated fluke eggs enter the deer's small intestine via the bile ducts and are passed out with the feces. Based on the lengths of time required for the various developmental events in the giant fluke life cycle, the commonest times that deer get infected are either in early spring or in late summer/early fall, when they graze near ponds and lakes.

A Trip to Europe

Giant liver flukes were first discovered and named in 1875 from specimens that came from an American elk or wapiti (*Cervus elaphus*) that died in a zoo in northern Italy. Despite first being found in Europe, *Fascioloides magna* was imported there from its natural home in North America, where its main mammalian hosts are primarily white-tailed deer, although elk, black-tailed deer (*Odocoileus hemionus*), and caribou (*Rangifer tarandus*) can also have egg-producing infections. Today, there are several "hot spots" of infection in North America that sustain giant liver flukes – usually areas associated with wetter conditions: around the Great Lakes (as the hunters in Michigan discovered), along the Gulf of Mexico coast and the southern Atlantic seaboard of the United States, in the North Pacific coastal area from northern California to southern Alaska, in the central Rocky Mountain trench in British Columbia, in south and central Alberta, in Montana, and in northern Quebec and Labrador.

In Europe, infections were reported sporadically from 1932 up to the 1990s, likely due to infected North American elk and white-tailed deer that had been imported to populate game parks. However, today the infection is common, especially in native red deer (*Cervus elaphus elaphus*) in the upper Danube River watershed. Besides red deer, fallow deer (*Dama dama*), and white-tailed deer are also core mammalian hosts. The infection pattern suggests that it took giant flukes about 150 years to become biologically acclimated to European conditions and to European snails, but now the flukes seem to be permanently established. Giant liver flukes serve as a lesson to us of the dangers of importing exotic wildlife.

Good Hosts and Bad Hosts

In North America, where the parasites evolved, many kinds of wild ruminant mammals can be infected with giant liver flukes (as well as domestic animals, rabbits, and guinea pigs experimentally). In normal final hosts (the hosts where mature adult parasites occur), the flukes live in thin-walled pseudocysts, which allow most eggs to escape into the mammal's intestinal tract. Most damage wrought by giant flukes tends to be caused by immature worms wandering through liver tissue, leaving tracts of destruction and hemorrhaging in their wake. There are few clinical signs of a host that is infected, but livers may become enlarged and have streaks, patches, and spots of black pigment. Normal host behavior does not seem to be affected by the parasites, but white-tailed deer body size and the number of antler points are reduced, which could impact deer reproduction since these features are used for attracting mates (see chapter 8). Usually, infection sizes of flukes are small enough (about five pseudocyst capsules per host) that there are no indicators of clinical disease.

However, that is not the case for other hosts. In domestic animals like cattle, sheep, and goats and in some wild cervids like moose, giant liver fluke infections can be serious and even deadly. These "aberrant" hosts may become slow, lethargic, and lose weight as disoriented immature flukes wander aimlessly in various organs. For moose, reproduction can be affected because of reduced antler quality and poor performance during the rut. Fluke eggs can block bile ducts, and thicker-walled pseudocysts can cause pressure on surrounding liver tissue, causing it to become rotten. Ruptured livers, general inflammation, and severe blood loss can result in death. Many of these hosts become "dead-ends" for the flukes since most eggs cannot get out into the environment to continue their life cycle.

Human Habitat Changes and Liver Flukes

In North America, the broken-up distribution where giant liver flukes occur today, with separated populations along the Gulf coast, in Labrador, around the Great Lakes, in the Rocky Mountains, and in the North Pacific, could mean that, in the past, these parasites had a larger range, and what we see today are reduced, relic populations. If so, some fluke populations may already be on their way to extinction, and those that persist may become specialized for local conditions and for the native snails and deer hosts. The giant liver flukes in Labrador and northern Quebec seem to be the most endangered because their caribou host population is drastically declining due to climate change and its effects on their northern habitat. With their main host gone, the flukes will likely be gone too.

However, the population of white-tailed deer in North America is expanding. Killing wolves and reducing other predators across much of the continent, and changing the landscape resulting in more deer-friendly habitat, has allowed white-tailed deer to become common, even in urban environments – I have had as many as four of them at a time eating perennials in my suburban backyard! Outside urban areas, making roads, trails, and cut lines for mining, oil and gas exploration and extraction, and clear-cutting of trees help deer to penetrate old-growth forests. Meanwhile, giant liver fluke intermediate hosts, lymnaeid snails, are common and widespread, and will do well if climate change creates wetter, warmer habitats with more ponds and sloughs. Climate models suggest that northern Canada will be warmer and wetter. Also, the successful invasion of Europe by giant flukes suggests that, over time, they can adapt and survive. This situation could mean that some of the isolated populations of flukes in North America are hanging in there and are on their way to becoming separate races or species. Time will tell.

Because ecological interactions and biotic community structure are inherently so complex and unpredictable, it is not a surprise that the effects on parasites and diseases of human-caused ecosystem changes and the rapidly changing climate are mostly a mystery. Perhaps an expanding white-tailed deer population, widespread and common snail hosts, and warmer (and in some cases wetter) conditions will permit the prevalence of giant flukes to increase. If so, we might see populations of vulnerable wild hosts like moose, and domesticated animals like sheep, be severely impacted. Perhaps soon, more giants will crawl among us.

5
Beetles and Beavers

In 1959, the "Father of Modern Ecology," a British biologist named George Evelyn Hutchinson, wrote a classic scientific paper entitled "Homage to Santa Rosalia or Why Are There So Many Kinds of Animals?" This paper was a must read for every ecology student. In it, Hutchinson presented a theoretical framework for the challenge of trying to explain community ecology: Are there any rules to predict what kinds of animals, and how many, live together in a particular ecosystem? What determines the capacity of a particular environment to support species? Six decades later, we are still trying to answer these questions. Today, we know that the ways in which energy and organic carbon are shared by organisms, the structure of the physical environment, and the nature of interactions among organisms are critical in explaining biodiversity, but we are still left with many questions.

An argument has been made that parasitism is the most successful biological strategy for getting resources that has evolved in the history of life and that it is responsible for more living organisms than any other method of harnessing energy – including photosynthesis. The ways that parasites have been able to invade hosts and to transform from free-living organisms into parasites have varied, and the strategy of becoming a parasite is not without danger. For instance, when parasites become irreversibly specialized to live with only one kind of host, if that host goes extinct, so will its parasites. Nevertheless, because parasites are so abundant and common, a different question than the one Hutchinson asked arises. Rather than ask why there are so many kinds of animals, we might ask: why are there not even more parasites?

Beetles

Our world has an amazing diversity of living things, from single-celled life such as bacteria, archaebacteria, and protists to the multicelled organisms that most of us are more familiar with – fungi, plants, and, of course, animals. But with all this incredible biological richness, one group of organisms has diversified to the point where biologists believe they make up more of the living world than any other organisms – beetles. There are more than 380,000 known species of beetles (close to one-quarter of all living species), but we probably have not made more than a dent in knowing the total number we share our planet with. This observation led famous British biologist J.B.S. Haldane, when theologians asked him what he concluded about the Creator from his study of nature, to say: "The Creator, if He exists, has an inordinate fondness for beetles."[1] This quotation speaks to the fact that there are more types of beetles than any other form of insect and more insects than any other kind of animal. Beetles belong to the order Coleoptera (*coleo* – sheath; *ptera* – wings). This name describes their most distinctive feature, front wings that are thick, leathery, or hard and form a sheath over the membranous hind wings. These unique front wings, called elytra, conceal and protect the membranous hind wings, which are used for flight and dispersal.

Another factor that has led to the success of beetles is their mouthparts. They are usually adapted for chewing, but beetle mandibles allow them to exploit many types of foods, ranging from soft fungus and plant parts to rock-hard wood. Also, beetles undergo complete metamorphosis, so larval stages use different habitats and foods than adults, thus avoiding intergenerational competition. Grub or caterpillar-like larvae hatch from eggs, pupate, and release the adults, which can disperse widely and colonize new habitats.

Beetles range in size from tiny *Scydosella musawasensis*, which lives in fungi in Nicaragua (only 0.3 mm or 0.01" long), to giant titan beetles (*Titanus giganteus*) in the Amazon, which are 550 times larger (167 mm or 6.6" long). Between these extremes, beetles' various characteristics have allowed them to be the most successful and diverse organisms on Earth, with the ability to live in almost every type of ecological habitat (the deep sea may be an exception). However, in spite of this incredible species richness, few beetles have taken up a parasitic lifestyle on mammals or on any other vertebrates.

Parasites and Parasite Wannabes

The few beetles that live with mammals occur mostly in South America and Australia; they often are not parasites but instead are harmless commensals or helpful mutualists.* For commensal beetles, only one life cycle stage (usually the adult) is found living with a mammal partner. For example, *Uroxys gorgon* and *Trichillum bradyporum* are scarab beetles found in the fur of sloths in Colombia and Costa Rica. Beetles lay eggs in sloth fecal deposits; after hatching, larvae eat sloth poop and then pupate. Adult beetles find a sloth and feed on algae and fungi that grow in sloth fur. Beetle feces may provide fertilizer that promotes algal growth, and algae in sloth fur are an important resource used by sloths as a salad to support their nutrition (see chapter 13). These beetles seem to have no negative effects on sloths, and by promoting sloth fur gardens, they may even be considered as mutualists rather than commensals. Although intimately associated with sloths, it seems clear that these beetles are not parasites.

Several kinds of adult rove beetles (staphylinids) occur in the fur of rodents and primates, usually around the anus, where they feed on fecal material. Among rove beetles, the ones that most closely show adaptations to be parasites are

---

* Commensal animals live in the nests, dens, or hair of mammals but do not harm them, while mutualists actively benefit their hosts.

six species in the genus *Macropocopris*, which are found on bandicoots and kangaroos in Australia. Only adult beetles occur in the fur, and they have well-developed tarsal claws (pincers on the ends of legs) that allow them to hold tightly to their hosts. Larvae never occur on mammals, and it is unclear what the adult beetles eat, so we really don't know if they are parasites or not.

Specialized beetle parasites that derive nutrition from host tissues occur in a family known as round fungus beetles (Leiodidae). Only 13 of 3800 species of round fungus beetles (about three-tenths of 1 percent) are parasites. These parasites all use rodents as hosts, especially rodents that are semi-aquatic. For example, one species, *Silphopsyllus desmanae*, is found on a rodent called the Russian desman (*Desmana moschata*) – a mammal that is about 20 cm (7.9") long and lives in river bank dens in Russia. The den entrances of these semi-aquatic rodents are below water, so the beetles are often submerged – they are little submarine parasites.

Beaver Beetles

The parasitic beetles that have diversified the most, however, are found on beavers – American beavers (*Castor canadensis*), European beavers (*Castor fiber*), and mountain beavers (Aplodontia). Mountain beavers (which are not really beavers but behave more like ground-dwelling muskrats and are related to squirrels) live in western North America in moist forests. The more familiar American and European beavers are the second largest rodents, next to capybaras, and are very

Figure 5. An adult beaver beetle, *Platypsyllus castoris* (dorsal view). Source: Gary Alpert.

adapted for aquatic life. Consequently, their beetle parasites have the most specializations and adaptations for living on hosts that are often submerged in cold water.

Beaver beetles occur in two genera, *Leptinillus* in North America and *Platypsyllus* in North America and Europe. A translation of "*Platypsyllus*" is "flat flea," which captures a sense of their parasitic lifestyle; however, it is a deceptive scientific name because these beetles are only very distantly related to true fleas (Fig. 5).

Look-Alikes That Are Not Related

While real fleas are laterally flattened (between left and right sides) so they can easily slip through hair, beaver beetles are dorso-ventrally flattened (we'll learn more about fleas in chapter 18). As a result, *Platypsyllus* beetles can press themselves tightly against their host's skin surface, under the dense fur coat where a layer of air is trapped when beavers are submerged. This arrangement not only keeps the beetles dry and provides air for them to breathe; it also keeps them warm so they can move around on their beaver host and feed.

The beetles, like fleas, have no hind wings – they cannot fly – and beaver beetles, like some fleas, have no eyes – they are blind. Dense beaver fur blocks most light, and beaver lodges are also dark, so eyes are of little value. Often, animals living in dark habitats lose their eyes over evolutionary time because eyes are so energetically expensive to grow and maintain.

Interestingly, some free-living round fungus beetles, which are relatives of the parasites, are specialized as cave dwellers and are also blind.

Beaver beetles have short, stubby sensory antennae, which are frequently tucked under the shield-shaped pronotum (the first thoracic segment) to keep the antennae protected from damage. After pupation in a beaver lodge, newly emerged adult beetles must find a host. Using their antennae, beetles are attracted to beaver odors and warmth, and they use their tarsal claws to attach. Locating a beaver by smell in the dark shouldn't be too difficult because beavers are famous for having castor glands – anal glands that produce a vanilla-like smell.

One structure that occurs on both fleas and beaver beetles is a row of stiff spines, like a comb, on the top side of the pronotum. This comb likely helps both fleas and beaver beetles to move in a forward direction through hair and to stay attached in the fur if the host tries to groom them. However, the spines mean that the beetles can only move in one direction – forward. Like fleas, beetles also have rows of stiff, backward-facing spines on their legs, which help them resist being dislodged by grooming or by normal host activity. All of the structural characteristics shared by fleas and beaver beetles provide a good example of convergent evolution – the separately evolved acquisition of similar characteristics by unrelated animals that share the same habitats, in this case as ectoparasites in dense mammal fur.

The Life and Times of a Beaver Beetle

Despite sharing some similar features, however, fleas and beaver beetles have very different lifestyles. Fleas most frequently occur off their hosts, in burrows or dens, and hop onto hosts for short times to suck blood. Fleas lay eggs in the host's nest, which hatch to produce worm-shaped larvae that scavenge on nest debris. In the nests, flea larvae pupate to form adults. Beaver beetles, by contrast, have a much more intimate association with their hosts, both as adults and as larvae.

Beetle larvae have piercing mouthparts, are found only on beavers (not in dens), and suck blood. Adult beetles have weak, flat mouthparts and do not suck blood. They feed on epidermal tissue, skin debris, and perhaps on skin secretions too. No one has ever observed infested beavers behaving as if they are bothered by the beetles, and no one has reported beavers using their split second toenails of their hind feet to try to scratch and dislodge either larvae or adults. Beaver beetles seem to be very cunning parasites that do not alarm their hosts to their presence.

After blood feeding, beetle larvae drop off their hosts in the dens and pupate in mud at the top of the lodge. The pupa is the only life cycle stage that does not occur on a beaver. Once newly emerged adult beetles get onto a beaver pelt, they nestle under the dense fur coat at the skin's surface and spend the rest of their lives there. Adult beetles can be found on beavers at any time of the year, while larvae are usually present only in the summer.

The Ecology of Beaver Beetles

Beaver beetles are not uncommon or rare. In 1963, ecologist Daniel Janzen studied beaver colonies in Minnesota and found that 82 percent of 38 beavers were infested, with an average of 51 beetles on each host (range 1–243). Janzen discovered that, rather than individual beavers being good or poor hosts, the environmental characteristics of each beaver lodge were more important for determining how heavily infested hosts became.

Lodges become the focal points of infestation by beetles, who get there primarily on two-year-old beavers that are establishing new colonies. For the beetles, life in a lodge is dangerous. Beetle stages that occur off the host (pupae) must deal with events like spring flooding, when many lodges get submerged for weeks. As floods recede, the interior of a lodge can be coated with a choking layer of silt or mud. Also, in late summer when water levels drop, beavers frequently avoid using dens because the entrances are well above the water level, so then beetles will have no access to hosts.

Nevertheless, adult *Platypsyllus* are well adapted to live with beavers. The parasites can survive and be active at cool temperatures (4°C or 39°F) but are readily killed by desiccation. These conditions, however, are rare because the beetles press themselves against their host's warm, moist skin. The beetles have become intimately tied to, and specialized for, living on beavers. They cannot survive on animals that do not have a dense layer of underfur that retains moisture. The transmission of beetles from host to host is by adults (not larvae), and close pelt-to-pelt contact is required, so the intimate family life of beavers keeps populations of parasitic beetles stable.

An Introduced Parasite

Today, we often hear stories about "foreign" organisms that have invaded unnatural ecosystems and, due to lack of predators or parasites, thrive and replace native flora and fauna. Invasive plants (blackberries in the Galapagos and Eurasian milfoil in North America) and invasive animals (zebra mussels in the Great Lakes and fire ants in the southern United States) are stubborn and expensive problems. One invader that has been allowed, however, is the *Platypsyllus* beetle on beavers.

In Scotland, the Eurasian beaver (*Castor fiber*) was driven to extinction in the sixteenth century by hunting and trapping. Attempting to reintroduce beavers, the Scottish government began a five-year scientific trial in 2008.[2] Starting in May 2009, 16 beavers from Norway were released in suitable habitat in western Scotland. Before release, the animals were determined to be free of diseases and pathogens, such as rabies and the potentially dangerous tapeworm, *Echinococcus multilocularis* (for more on tapeworms, see chapter 11). However, one parasite that was inadvertently reintroduced to Scotland was the beaver beetle, *Platypsyllus castoris*. Unlike most invasive species, however, there is little concern that this beetle will switch to other hosts since they can only live on beavers, nor is it likely that they will become a serious pathogen of the reintroduced mammals. An indicator of the beetle's harmlessness is that the beaver population in Scotland has grown from the original 16 immigrants to more than 150 (making the project one of the most successful reintroduction conservation programs ever). The specialized morphology and unique life habits of the beetles means that they are extremely unlikely to infest any other hosts.

A Biological Puzzle

Beetles are the most biodiverse animals on Earth and have exploited almost all ecological niches, so it is surprising that so few have acquired a parasitic lifestyle. The few that have become parasites seem to have gone from being commensals in moist mammal nests or dens onto the fur of semi-aquatic rodents like beavers. But why these hosts, in particular?

Beavers have a very specialized natural history. They have plush, dense fur, live in colder habitats, and are frequently submerged. This lifestyle means that they are free of most ectoparasites, including lice, fleas, and ticks (*Ixodes banksi* is the only one reported), but they are infested by mites. Like the beetles, beaver mites have become very specialized – so specialized that they show niche partitioning on their hosts. Different species are found only on the head and neck, others on the abdomen, on the back, or on the front or hind legs. However, it is not clear whether these mites are parasites or ectocommensals. Like feather mites that occur on birds, beaver mites may be mutualists by helping beavers keep their fur clean and in good order since mites eat dead skin cells, debris, and fungal spores in the pelts.

This beaver lifestyle might explain why beetles have become specialized parasites of beavers. Beaver beetles discovered a challenging (frequently being submerged in cold water) but unexploited niche. They found a humid, warm microclimate with a trapped layer of air for respiration, plentiful food (maybe, for adult beetles, food includes resident beaver mites in addition to skin cells and debris), and no competitors. The beetles were likely "preadapted" (or, maybe a better term, "predisposed," because evolution cannot see ahead) to survive in this unique habitat, and perhaps the parasites' ancestors came into frequent close contact with their future hosts in dens. Today, the beetles have so many specializations for infesting beavers that they have gone down an evolutionary one-way street. If beavers go extinct, so too will the parasites. Beaver beetles seem to cause little damage to their hosts, who largely ignore them. They may represent an evolutionary experiment in specialization, but it seems unlikely to lead to much divergence, unlike the pattern we see for most free-living beetles. Perhaps if and when beavers speciate (but their rate of evolution appears to be very slow), new species of parasitic beetles will form.

When naturalists started collecting insects and found the overwhelming biodiversity of beetles, they wondered why these animals have become so numerous and diverse. One hypothesis suggested that, as flowering plants diversified, so too did beetles. Beetles could utilize many different parts of plants (pollen, nectar, leaves, stems, roots, seeds, bark) and specialize on each different kind of plant, which permitted their adaptive radiation and massive expansion into different and new varieties. Another idea, however, is that beetles are uniquely built to avoid extinction – rather than unduly increasing their numbers, they simply have not gone extinct as frequently as other organisms. Whichever of these ideas is the reason for the wonderful biodiversity of beetles, or even if both happened, neither explanation answers the puzzle of why more beetles have not taken up a parasitic existence on mammals. Perhaps Haldane's Creator, besides being inordinately fond of beetles, held a special kindness for mammals as well.

6

Stranded Whales: A Fluke Accident?

Although more than 70 percent of our planet is covered in oceans, our lack of knowledge about the life they contain is so great that they might almost be on Mars. Our ignorance is alarming, yet one thing we do know is that we are causing great changes to the world's seas. Ocean temperatures and sea levels are increasing; atmospheric, man-made carbon dioxide is being absorbed in seawater, causing acidification; and increasing levels of pollution (including massive amounts of plastics) can be seen everywhere. Because of our ignorance, we have no idea what the consequences of these changes will be on marine organisms, but one kind of animal is likely to quickly reflect the stresses we are putting on our seas – cetaceans (dolphins, porpoises, and whales). Like all mammals, cetaceans are infected by a variety of parasites. How changes to marine ecosystems will affect the

72　Fur, Fleas, and Flukes

Figure 6. Long-finned pilot whales stranded at Farewell Spit, New Zealand (2017). Source: Gary Webber / Alamy Stock Photo.

balance between cetaceans and their parasites is a puzzle, but evidence from mass strandings of whales should set off an alarm.

A Mass Stranding Event

On the morning of February 10, 2017, and during the next two days, more than 600 long-finned pilot whales (*Globicephala melas*) floundered on the beaches of a small crescent of land called Farewell Spit on the South Island of New Zealand (Fig. 6).[1] As often happens during whale stranding events, many compassionate volunteers worked tirelessly to return still-living whales to the ocean, but more than 300 whales were already dead, and some that were rescued beached themselves again. This mass stranding event was large but not unprecedented. Around 1920, 1000 whales were stranded off the east coast of New Zealand, and in 1985, 450 whales beached themselves near Auckland. What could possibly drive whales to commit mass suicide with an impulse so forceful that those that are saved repeat the attempt?

Pilot Whale Biology

Pilot whales, like orcas (killer "whales") are not really whales. They are one of the largest types of dolphins (family Delphinidae). Unlike true whales, dolphins have many conical, peg-shaped teeth, although pilot whales have fewer (48) than most delphinids. They also feature large and elaborate cranial sinuses around their ears so that their well-isolated ear bones give them excellent echolocation ability, which in turn allows them to feed on fast-moving prey like fish, and especially cephalopods including squid, cuttlefish, and octopuses. Their echolocation ability is enhanced versus other cetaceans in part because they have asymmetrically shaped premaxillary (front of the upper jaw) bones. The asymmetrical jaws are thought to work like the asymmetrical ears of owls; they permit acute, directional detection of sounds.

Pilot whales are long-lived (about 40 years for males, 60 for females) and have delayed maturity. Males mature more slowly than females (12 years for long-finned males, 8 for females), and once mature, pilot whales have seasonal mating. Females produce one calf every few years; mothers nurse their young with very calorie-rich milk for three years, and in most pods there are many post-reproductive females (matrons).

Pilot Whale Tweets

Like orcas, pilot whales are very social animals, living in pods of up to 90 members. Most individuals in a pod are genetically related to mothers and grandmothers, and pods have whales of all ages and sexes, although females usually outnumber males. Pod membership is stable – individuals spend their entire lives with the same animals. Males, however, will breed outside their family group, which promotes genetic diversity. Mating is polygynous (males mate with more than one female) and occurs when pods periodically come together to form large aggregations. As a result, pilot whales are organized for males to stay with female kin in pods but to mate with unrelated females from other pods.

With such a structured and intricate social system, and with animals that live together for long time periods, it is not surprising that communication among pod members plays a crucial role in the lives of pilot whales. Besides navigating and hunting using echolocation, pilot whales have complex behaviors that are facilitated and coordinated by vocal communications. In fact, pilot whales were given that name because pods tend to follow one or two leaders (pilots), with the animals all keeping in contact using calls. Whales keep in touch with each other constantly, and each pod likely has its own unique, group-specific set of calls. Sonic analysis of calls shows that pilot whales produce many kinds of vocalizations, including simple and complex whistles and pulsed sounds – tweets. The types of sounds correspond to various behaviors – restful "milling," active feeding, courting, mating, predator alerts, and so on. The types of vocalizations change, however, depending on how concentrated pod members are. Individual pilot whales have their own signature whistles, which change depending on an animal's mood (for instance, if they are threatened, or hungry). It is likely that each pilot whale in a pod can recognize individuals in the pod by their unique vocalizations. Might the complex social structure, behaviors, and vocalizations of pilot whales contribute to mass stranding incidents, such as those in New Zealand?

Pilot whales are among cetacean species that are most frequently reported in mass and single-animal strandings. Although strandings have been reported for years, the cause(s) for these heart-wrenching events are still mysterious. Storms and seabed topography have been proposed as factors: shallow sloped sites may confuse echolocation, and different stranding events are sometimes seen at the same locations. Perhaps whales follow large schools of prey, or are hazed by predators like orcas, into shallow areas. Even human activity has been implicated in strandings. Low frequency military ship sonar and the thuds of large ship propellers may cause acoustic pollution, which damages whale ears and air sinuses and disrupts their navigation. However, another factor that has been tied to mass and single-animal strandings is parasites.

Fluke Accidents?

Eggs and adult flatworm flukes (phylum Platyhelminthes, class Trematoda), belonging to the genus *Nasitrema*, are often found in delphinids with melons (bulbous chambers filled with fats, oils, and waxes) in their air sinuses. Although melon function is not completely understood, it likely acts as a resonating chamber, allowing pilot whales to project and direct vocalizations, and enhances their navigation and communication skills. *Nasitrema globicephalae* occurs in long-finned pilot whales and usually lives in air sinuses, but sometimes they occur in the animals' brains. The air sinuses are cavities in the upper head, used for sound production, and in the lower head, used for hearing by isolating the ears to improve directional sound detection. Both the sinuses and the melon of whales are not completely enclosed by bone, so they are subject to pressure changes when the animals dive. In addition, they are richly supplied with blood vessels and likely play a role in regulating air to prevent whales from suffering "the bends" (bubbles of nitrogen gas entering the blood and damaging kidneys, lungs, liver, and the central nervous system).

Nasitrema flukes are often associated with single-animal strandings. Affected whales have a loss of equilibrium and appear disoriented. Adult flukes most often occur in small numbers (less than 10), but infections of 100 to 200 worms have been found. Adult *Nasitrema* flukes are small (about 1.5–2 cm or 0.6–0.8" long), but their bodies are covered in numerous spines that likely cause mechanical damage and inflammation to sinus linings. No one knows what the flukes feed on. However, they are attached to the mucus-secreting linings of the sinuses, so perhaps, in addition to the spines being used for attachment, the irritation caused by them may result in more mucus production, which the flukes would eat. Because the sinuses are also highly vascularized, the flukes may induce some hemorrhaging, so they could potentially also feed on blood.

The fluke eggs are golden-brown and triangular, and they get from sinuses into the lungs and can cause respiratory congestion (mucus buildup). As a congested whale tries to get its blowhole out of water to breathe, it may lodge itself on a shallow beach or sandbar and become stranded. The worst pathology, however, is caused when fluke adults and eggs occur in the brains of whales. Here, they cause inflammation of the meningeal linings and brain tissue damage in the cerebral hemispheres. In the brains of infected animals, the flukes cause disruption of white and gray matter, tissue death, and the formation of cavities in the cerebral hemispheres. One reported case in a dolphin found damage to the vestibulocochlear nerve, which affects equilibrium and echolocation.[2] If the facial nerve was similarly damaged, then besides other problems, the host's ability to regulate gases during deep dives would also be affected.

Just as for their delphinid hosts, infection of the brain by flukes is not good for *Nasitrema* either – the brain is a dead-end site with no way for eggs to escape the host. Trapped eggs will be attacked by the host's immune system and will cause a lengthy, "smoldering" inflammation reaction, resulting in central nervous system damage and tissue death.

Besides pilot whales, *Nasitrema* has been reported in 11 other species of whales and dolphins.

Despite their frequency and the damage they can cause in hosts, we do not know the life cycle of any of the 10 known species of *Nasitrema* nor, for that matter, for any related flukes (Brachycladiids). Based on the life cycles of most other flukes, it is likely that gastropod snails act as first intermediate hosts, and fluke infective stages (metacercariae) probably encyst on fish or possibly on invertebrates such as squid. If so, the parasites would be transmitted when definitive hosts (whales) eat fish and squid. Studying parasites and diseases of marine mammals is very challenging because these studies are expensive, experimental infections are almost impossible, and there are ethical constraints.

Flukes and Mass Strandings

Besides individual cases of disease and single-animal strandings, could *Nasitrema* flukes also cause mass stranding events like the Farewell Spit incident? Even if only a few whales in a pod were heavily infected, but one or more of these were leaders (pilots), infection could result in an entire pod following a sick, disoriented leader into dangerous shallow water. Also, if the sick, stranded animals gave distress calls, those calls could drive healthier pod members to continuously re-strand in futile attempts to aid the sick pod member(s). In the mass stranding event in New Zealand, the characteristic triangular-shaped eggs of *Nasitrema* were often found in brain tissues of the pilot whales. Perhaps whale echolocation and communication in the particularly dangerous shallows of Farewell Spit were affected by the flukes. Because the flukes occur in air sinuses, which play a role in gas regulation during diving, fluke damage might also cause decompression sickness. Some postmortem examinations of mass stranded whales have found gas and fat bubbles (embolisms) and hemorrhages in many whale tissues,[3] but these studies did not look for the co-occurrence of *Nasitrema* flukes.

Despite many studies of whale stranding events, we still are not certain of the causative role of *Nasitrema* or other parasites. It could be that whales affected by additional problems, like sonar damage, malnutrition, or toxic pollution, have weakened resistance and get infected with more parasites, or these factors could cause parasites to undergo aberrant migrations from sinuses into whale brains. Based on studies of other parasites in hosts,[4] we know that, in heavy infections, some worms may be forced by competition into sites in the host where they usually do not occur. Although we know little about the basic biology of *Nasitrema* and its life cycle, this type of knowledge is crucial to warn us if infections and diseases in marine mammals are increasing or if they are related to physical and chemical changes that are occurring in our oceans, such as warming or acidification. We know that mass stranding events have been happening for many years, but only more studies will help us to know if they are becoming more frequent and how (or if) parasites play a role in these sad events.

Today, like terrestrial ecosystems and our atmosphere, oceans are changing rapidly. Warmer, more acidic water, polluted with masses of plastics, is the new normal. Sadly, we know less about our oceans, and the life they support, than about any other ecosystem.

Unfortunately, we currently have so little baseline information that it is impossible to know how environmental changes will affect marine parasites and diseases, so our hopes of understanding, mitigating, and preventing deaths of marine mammals by events such as mass strandings seem far off.

7
How the Zebra Got Its Stripes

Mammals can be boring – at least with regard to their colors. While many birds, like peacocks, hummingbirds, and tanagers, have lavish plumage dyed in any color you can dream of, and poison dart frogs are splotched in gaudy blues, yellows, neon greens, and reds, mammal fur is usually black, gray, brown, reddish-brown, blonde, or white. Very few mammals sport bright colors. Socially dominant male mandrills (*Mandrillus sphinx*), which are large, old world monkeys, have bare skin patches on their buttocks and faces that can display bright blue, purple, and red (due to skin cells that have strings of collagen protein that reflect these iridescent colors). Blue whales (*Balaenoptera musculus*) are grayish-blue (they look brighter blue under water) and have sulfur-yellow bellies due to a coating of algae (diatoms). Sloths (*Bradypus* and *Choloepus*) have green fur due to algae that grows in the shafts of their hair. But brightly colored mammals are the exception – not the rule. However, what mammals lack in coat colors they make up for with a great array of patterns, including solids, spots, splotches, masks, eye patches, and of course, stripes, like those on zebras.

Evolutionary biologists sometimes ask the question "why." For instance, is there an evolutionary answer for why leopards have spots or zebras have stripes? Biologists propose different possible selection pressures that would favor a particular adaptive characteristic; they then collect data or conduct experiments that try to eliminate the possible role of these selection pressures until one or two gain the most supporting evidence. Evolutionary biologists must show that a characteristic they are interested in (such as zebra stripes) is genetically inherited, so natural selection can modify it, and also that the selection pressure they think is responsible has been present long enough and intense enough to cause the feature to become genetically fixed in a population. Could parasites possibly be a selection pressure that would result in a coat pattern becoming a characteristic of a mammal? Or, to put the question another way, why do zebras have stripes?

Different Explanations

There are many legends about how zebras got their stripes. According to the San bushmen of the Kalahari Desert, long ago when animals first appeared in Africa, the weather was hot and water was scarce. A pool was guarded by a rambunctious baboon that forbade other animals from drinking. One day, a zebra and its son came to drink. The baboon jumped at the zebras and barked that the pool was his alone. The zebra's son, who was very thirsty, shouted back that the water belonged to everyone. A fierce fight resulted that kicked up a large cloud of dust. Finally, the zebra got in a mighty kick, sending the baboon high into the surrounding rocks. The baboon landed on the rocks so hard that the hair on his backside came off – that's why baboons have bare patches there today. Baboons remained in the rocks and to this day hold up their tails to ease the pain of their bald bottoms. The exhausted zebra, blinded by the dust, staggered near the baboon's campfire, which scorched him and left black marks across his white fur. The shock scared the zebra, and off he galloped onto the savannah plains, where he stayed forever after.

After diverging from horses (*Equus caballus*) about 3.8 million years ago, at least nine species of zebras evolved, including quaggas (*Equus quagga*) and wild asses (*Equus asinus*), which ranged over all of Africa. Today, only three species remain – Burchell's or plains zebra (*Equus burchelli*), Grevy's zebra (*E. grevyi*), and mountain zebra (*E. zebra*) – and all have greatly reduced distributions from those they historically held. Besides how, why, and when zebras got their stripes, evolutionary biologists ask many other questions about the beautiful striped coats zebras have. For instance, are zebras white with black stripes or black with white stripes? Do the stripes serve any purpose? If the stripes do have a function, is it the same as for other striped animals?

Early biologists thought that zebras were white with black stripes because the bellies of zebras are white. The zebra pattern forms during fetal development. Different numbers, widths, and positions of stripes help to define the three different species. Black stripes arise from mature cells called melanocytes that deposit melanin pigment into hairs, but in some melanocytes, pigment deposition is genetically switched off, resulting in white fur. The unique pattern for each zebra species is likely affected by the timing of when (and if) melanocytes make melanin, but the black and white patterns they have are genetically inherited.

If pigment cells are positioned and turned on in early development, stripes are wider and less numerous; if turned on later in development when the embryo is bigger, more numerous, thinner stripes occur. When the fur of a zebra is removed, the thin, underlying skin is dark, so some argue that zebras are black with white stripes. These are the proximate mechanisms (things that happen during the life of an organism) that explain zebra stripes, but what are the ultimate, evolutionary reasons?

Selection Pressures for Stripes

The controversial question of the possible function of zebra stripes intrigued the co-discoverers of evolution by natural selection, Charles Darwin and Alfred Russel Wallace. Darwin thought that the bold stripes played no role in deceiving predators:

> The zebra is conspicuously striped, and stripes on the open plains of South Africa cannot afford any protection. Burchell

in describing a herd says, "their sleek ribs glistened in the sun, and the brightness and regularity of their striped coats presented a picture of extraordinary beauty."[1]

However, Alfred Russel Wallace argued that the patterns help animals to blend into the background at dusk:

> It may be thought that such extremely conspicuous markings as those of the zebra would be a great danger in a country abounding with lions, leopards and other beasts of prey; but it is not so … It is in the evening, or on moonlight nights, when they go to drink, that they are chiefly exposed to attack … [I]n twilight they are not at all conspicuous, the stripes of white and black so merging together into a gray tint that it is very difficult to see them at a little distance.[2]

Today, there are several ideas for the possible function of zebra stripes. As debated by Darwin and Wallace, the stripes may play some role in camouflage or confusion of predators. Other ideas suggest that the stripes are used by zebras as signs to confer information about the type or social status of the animal, or for thermoregulation (to help zebras stay at a comfortable body temperature), or perhaps to deter or confuse parasites.

How might a zebra pattern affect predators? Some suggest that the vertical stripes make zebras look big or disrupt an animal's outline, while others think they provide camouflage, especially in habitats with strong, vertically patterned vegetation such as tall grass or brush. These ideas, however, are unlikely. In a 2016 study, sensory ecologist Amanda

Melin and her colleagues used field data, digital images, and psychophysical methods to show that zebra stripes did not provide protection from lions (the primary predator of zebras) or hyenas in any kind of lighting conditions in either plains or brush habitats. Also, the stripes did not confuse predators or cause a "dazzle effect." Naturalists also note that lions kill proportionally more zebras (regardless of herd size) than other prey, so if stripes are for camouflage, they are not working very well. Nevertheless, it's possible that lions would kill even greater numbers of zebras if zebras had no stripes.

Could stripes be used as signals for zebra communication? Zebra herds include loose social groups that are leaderless, so no social dominance cues are given by stripes. Zebras don't use the stripes to recognize members of their own species – today, there is little overlap in the distributions of the three species, so this function isn't needed. Behavioral studies of zebras[3] find no evidence that the stripes are used for individual recognition or for signaling mating quality or health status, and the stripes don't elicit grooming by another zebra or facilitate social bonding between mares and their foals. Zebra stripes are not involved in communication.

Could zebra stripes help animals to thermoregulate? Because black and white stripes absorb and reflect solar radiation differently, thick black stripes will heat more than thick white stripes. This difference will create air eddies, which might cause cooling. Thus, zebras with thicker, darker stripes should occur in the hotter, more tropical parts of Africa to create stronger air currents. Several observations and studies support the idea. For instance, plains zebras exhibit a gradual change in stripe thickness across their geographic range – zebras closer to the equator are darker than zebras further south.[4] This difference suggests that an environmental factor may be involved in selecting stripes. Studies have found that predators and camouflage can both be ruled out as factors affecting striping, and while those studies showed that environmental factors including temperature and humidity could potentially be involved, other observations don't support the idea that stripes evolved as a thermoregulatory mechanism.

For instance, surface temperatures of plains zebras in mixed herds with other hoofed mammals are no cooler than other animals. The black stripes of plains zebras are larger than the white, so more heat would be absorbed; the fur around the nostrils and mouths of zebras is also dark, so it would heat the air that they use for respiration. The surface of a zebra's back has a dark stripe down the middle; this surface is exposed to direct sunlight, so if cooling were a problem, this area should not be black. Also, the dusty plains habitat is often windy, so any cooling air eddies that might be created by the stripes would be disrupted by the breeze, and when zebras move, they would also disrupt cooling air currents. There is also experimental evidence that stripes do not cool zebras.[5] Finally, other hoofed grazers that share the same habitat, like wildebeests, eland, and topis, are not striped. Thus, because of little or only mixed support for any of these ideas, biologists have looked at another environmental feature that might select for stripes – parasites.

Parasites and Stripes: Bots and Other Flies

Zebras, like all wild mammals, must deal with a variety of internal and external parasites. Besides ectoparasites like ticks and lice, zebras are attacked by several different types of flies. True flies (order Diptera) are insects that have become highly adapted for flight. Instead of the typical pattern for insects of having two pairs of wings, true flies have changed one of the wing pairs into a navigational device called a haltere. This knob-like structure gives flies information about their position in the air (yaw and pitch), similar to a gyroscope on an airplane. In addition, flies have large, compound eyes that give them visual acuity but are especially good at detecting motion. As a result, true flies are skilled aerial acrobats that can travel fast through the air or hover over a host.

Two kinds of bot flies (Oestridae and Gasterophilidae) look similar to stout house flies. Both infect zebras when female flies glue eggs onto hairs of zebra legs, throats, around the lips, or in the nasal passages directly. Nasal bots (Oestridae) develop from larvae (maggots) to pupae in the nasal passages of zebras and are then expelled (often by sneezing) onto the ground, where they form adults. Stomach bot (Gasterophilidae) eggs are also glued to zebra hairs and are stimulated to hatch by the licking and grooming of their host. Eggs develop into larvae, which are swallowed and attach in the mouth or pharynx (the cavity behind the mouth that leads to the esophagus). The larvae then migrate or are swallowed into the zebra's stomach, where they attach to the stomach wall using sharp, recurved barbed hooks to anchor themselves in the host. After feeding and growing in the stomach, the bots get passed out with the zebras' manure, pupate on the ground, and a new crop of adult flies emerge. Stomach bots are very common in zebras – often 100 percent of animals in a herd are infected, and parasite populations (bots in the stomach) can reach hundreds. Although often asymptomatic, stomach bots can cause inflammation of the stomach lining and lead to ulcers (open sores). Because bots are so common in zebras, it does not seem likely that stripes affect the ability of these parasitic flies to find and lay eggs on their host's fur.

Tsetse Flies

Other flies that are important in zebra habitat are tsetse flies (*Glossina* spp.; Fig. 7). Tsetse flies are related to our common house fly, but they live only in tropical sub-Saharan Africa and are blood feeders. Unlike mosquitoes that can take blood meals without a host knowing, tsetse fly mouthparts have tiny saw-blades on them, and their bites are painful. Besides being an irritant, however, tsetse bites are very dangerous because they transmit a single-celled parasite called *Trypanosoma brucei*, which causes a disease called nagana (the animal equivalent of African sleeping sickness in humans). Infected wild mammals, such as wildebeests, seem to develop some resistance to the infection, but domestic animals such as cattle, sheep, and goats become emaciated, lose weight and condition, and frequently die. Wild mammals with trypanosomes act as a reservoir of infection

Figure 7. Zebra stripes likely deter biting flies, such as tabanids (top left) and tsetse flies (top right). Sources: top left: Henrik Larsson / stock.adobe.com; top right: Dorling Kindersley / UIG / Bridgeman Images; bottom: Daniele_K / Pixabay.

(a constant pool of infected animals) that leads to infections in livestock. Nagana is so pervasive and debilitating that it has impeded ranching wherever tsetse flies are common.

However, tsetse flies do not seem to be a major concern for zebras. Studies that have analyzed blood meals of the flies found that plains zebras in Serengeti National Park in Tanzania were seldom fed upon, even when they were among the most common animals available.[6] Also, there are few reports of the trypanosomes that cause nagana in the blood of zebras. We'll need to continue our search before we find a likely parasitic candidate for causing zebra stripes.

Tabanid Horse Flies

Tabanid flies are called "horse flies" for a good reason – they can be an important, irritating, and painful problem for any kind of equid, including zebras. Female tabanid flies are large and need to have a blood meal to produce eggs. Their bite involves stabbing and cutting a zebra with their mouthparts by slicing the host's thin skin using scissor-like movements of serrated, knife-like mandibles and maxillae. The flies spit saliva into the wound, which includes anticoagulant proteins that slow clotting, so a pool of blood forms at the bite. Horse fly mouthparts also have a spongy pad they use to sop up the pooled blood. One zebra tabanid can take 0.5 ml (or about a tenth of a teaspoon) of blood in a single meal, and female flies require several meals to produce eggs. Blood loss from many bites can result in anemia, which results in weakness and tiredness. I've never been bitten by a horse fly myself, but based on reports of others, the bites are very painful.

Horse flies are a serious irritant for zebras. Zebras alter their behavior to try to avoid flies and expend energy to swat flies with their tails, shimmy their skin, roll in dust, and move to windy, higher ground. Besides causing irritation and blood loss, tabanids also spread many infectious diseases to zebras, including viruses (equine infectious anemia virus), bacteria (anthrax), protozoans (trypanosomes, but not *T. brucei*), and even animal parasites such as roundworms (*Loa loa*). Because it is so painful, tabanid blood feeding is often interrupted by a host, so zebra flies frequently move to another member of the herd, which promotes the spread of these diseases.

Tabanid flies find hosts by detecting movement, body heat, and breath (expelled carbon dioxide), and they are visually attracted by color, pattern, and contrast with the background. They tend to attack the legs of zebras, where they are more difficult to swat, and zebras tend to have more stripes on their legs than elsewhere. This observation led researchers to wonder if zebra stripes evolved to reduce fly attacks. Instead of suggesting that the temperature and humidity relationship to stripes was due to thermoregulation, it was pointed out that these same environmental factors are also related to tsetse and horse fly densities. For instance, although hungry adult female horse flies prefer drier open areas, eggs are deposited in wetter areas, and larvae develop in humid, muddy locations, which necessarily affects the fly density in those different environmental conditions. In addition, the degree of zebra striping and the geographic overlap with higher densities of biting flies showed that, in areas with lots of flies, zebras had the most stripes. Could zebra stripes reduce attacks by biting flies, especially tsetse and horse flies?

This idea was tested experimentally.[7] Insect traps of different colors and patterns showed that vertically striped traps collected fewer biting flies. Also, when horses that were pastured together were draped with solid white, solid dark, solid beige, or black and white striped coats, fewer flies attacked striped horses. When the researchers themselves wore solid or horizontally striped clothes, they attracted more biting flies than they did wearing vertical stripes. Ecologists have shown that moving targets with stripes attracted half the number of tsetse flies and one-third the number of horse flies as did solid black or white targets.[8] These studies also showed that fewer flies were attracted when stripes were closer together.

Tabanids find hosts using chemical cues, but they use visual cues to land. The visual acuity of the insect compound eye is poor compared to humans, but they are better than us at detecting movement, which is why it is so difficult to swat a fly. Object detection by biting flies seems to depend on size, shape, color, contrast, and pattern, and for most flies, dark colors are preferred. Horse flies and tsetse flies can find host targets regardless of the coat pattern, but they have trouble landing on striped surfaces – the light and dark stripes of a zebra's coat reflect very different polarizations that disrupt the ability of tabanids to land. Also, zebra stripes may make zebras unattractive to horseflies because they confuse the tabanids' ability to detect the heat of blood vessels in the skin.[9] Ironically, the arguments made long ago by Wallace for zebra striping hiding zebras from (and confusing) predators are, in a sense, both true – except, unbeknownst to him, the predators of importance were not lions, leopards, and hyenas but micropredators like tsetse flies and horse flies.

Agricultural researchers in Japan decided to put the fly-protective function of zebra stripes to a practical test by painting zebra stripes on Japanese black cattle.[10] As controls, they left some cattle unpainted, and they also painted black stripes on others to test if the process of painting might have an effect. The stripes had an amazing effect – the total number of flies (including horse flies) that landed on the "zebra" cattle was less than half those that went to the controls, and the striped cattle also spent significantly less time using fly-repelling behaviors. The researchers pointed out that painting cows could reduce the need for ranchers to use expensive and environmentally damaging pesticides – as long as you don't mind funny-looking cows.

Although several studies have implicated biting flies and the infections they transmit as a major selection pressure that led to the evolution of zebra stripes, more research needs to be done to make the relationship clearer.

However, one thing is certain: when Darwin and Wallace speculated on the reason for zebra stripes, they opened a huge Pandora's box of questions that will keep evolutionary biologists busy for years to come. We do know that parasites can have drastic effects on the health and condition of individual hosts and that they can also affect the ecology and distribution of hosts. Now it also seems likely that some of the physical features of wild mammals, such as the amounts, distribution, composition, colors of fur, and color patterns that are so representative of some mammals like zebras, could be due to parasites.

8
Ornaments and Parasites

Using "unnatural" selection, big game hunters and poachers often selectively kill mammals that are bigger than average or have larger and more elaborate or symmetrical trophies or ornaments (antlers, horns, tusks). The prized mammals tend to be older, better fed, in good condition, and socially dominant. Frequently, these trophy animals reproduce more in the population and pass on more copies of their genes to the next generation (or at least they would if not killed). By contrast, many TV nature shows suggest that, unlike trophy hunters and poachers, predators use natural selection to kill the weakest, sickliest members of the population. Do parasites act like predators? Are parasitized animals those with the poorest quality ornaments? These kinds of questions prompted evolutionary ecologists to propose what at first seems like a simple idea in order to explain the relationship between parasites and sexual display ornaments – sexual ornaments signal parasite infections.

Figure 8. The magnificent antlers of Irish elk, *Megaloceros giganteus*. Source: iStock.com / ZU_09.

A Magnificent Mammal

An Irish farmer, while digging peat in a bog in the Shannon Basin east of Dublin, was surprised when his spade struck something hard just beneath the surface. As he uncovered it, he recognized that the object was a skull and antlers of an extinct giant – an Irish elk (*Megaloceros giganteus*). When fully revealed, the massive antlers were over 3.5 m (11.5') long and weighed 40 kg (88 lbs)! Surprisingly, despite their common name, the magnificent mammals that bore these impressive structures were neither elk nor confined to Ireland.

The antlers in fact belonged to giant deer that once lived all over Europe and as far south as North Africa. When sea levels dropped during the Pleistocene, the mammals colonized Great Britain and Ireland, where some specimens were preserved. Based on remains that have been found, Irish elk were over 2 m (6.6') tall and weighed about 600 kg (1320 lbs). As massive as they were, Irish elk were not the largest deer known. That title belongs to North American stag moose (*Cervalces scotti*), which stood 2.5 m (8.2') tall and weighed over 700 kg (1540 lbs). However, Irish elk antlers are by far the largest antlers known from any mammal (Fig. 8).

Antlers and Sexual Selection

During the rut, male Irish elk used these elaborate, heavy ornaments to attract females and to spar with other males. Incredibly, the antlers were shed annually, so Irish elk had to use their metabolic energy and precious nutrients to build a new, larger rack each year. Antler size is a heritable character, passed from father to son, so natural selection should act to penalize organisms that don't have features that promote their survival and successful reproduction. After discoveries of Irish elk and the realization that these mammals were extinct, some early naturalists argued that the reason for the demise of Irish elk was because the massive antlers were "maladaptive" – the antlers were so large and so cumbersome that they led to the species' extinction. Some zoologists even argued that antler size demonstrated "run-away" evolution due to directional selection – antlers constantly got larger and larger until at last they became unsupportable structures.

In 1871, Charles Darwin had a different idea. He reasoned that supposedly maladaptive ornaments, like male peacock tails and Irish elk antlers, confer a reproductive advantage. These "secondary" sexual characters may not help in the survival of the male that carries them, but they play an important role in reproductive success. Males with larger ornaments may gain access to more females, father more offspring, and therefore leave more copies of their genes in the next generation. This kind of natural selection (Darwin called it sexual selection) would favor male elk with larger antlers, even though there could be a heavy cost to the elk's survival. As long as the animal with large antlers lives long enough to reproduce often, and if the structures are attractive sexual display ornaments that allow male mammals to have more reproductive opportunities, these ornaments should then promote greater overall Darwinian fitness and be favored by natural selection.

The Hamilton-Zuk Hypothesis

Because ornament size and symmetry are partly determined by genes, female mammals may choose mates with better ornaments so they will pass these genes to their sons, who will then have more reproductive opportunities – the "sexy son" hypothesis. How might this pattern have an effect on parasite infections? If immunity or resistance to parasites and diseases is also heritable, then ornaments may truthfully inform a female that she is mating with a healthy, less-parasitized mate, so she will not be exposed to his pathogens during mating and will have offspring that are more resistant to infections. The physical condition of an ornament is a truthful indicator because males can't fake it like they could if behavioral signals were used. Of course, this idea implies that parasites can affect the size or structure of ornaments. The theory that ornaments and parasites are related was developed by evolutionary biologists Bill Hamilton and Marlene Zuk in 1982, and because it is such a fascinating idea, their hypothesis stimulated lots of research, particularly with birds and insects.

Testing the Hamilton-Zuk idea in wild mammals at first seems easy – just compare the parasite loads of males that have well-developed ornaments versus those with poorer quality structures. Unfortunately, simple correlations like that may not really explain what's going on. For instance, it may be necessary to show how (and when) parasites affect size or symmetry of ornaments, that immunity to parasites is heritable, and that males with fewer parasites sire more offspring. As a result, the commonsense idea that female mammals get a health status report when they assess male ornaments is not that easy to validate.

Tooting Your Own Horn: Other Ornaments

In order to test the Hamilton-Zuk idea with mammals, researchers have studied antlers, as well as other structures, that are differently developed in males versus females – such as horns and tusks.[1] Many people confuse antlers with horns. Antlers are composed of deciduous bone and are branched. They occur on deer, elk, moose, and caribou, usually only on males (with the exception of caribou). Regulated by hormones, antlers are shed and regrow annually – stimulated by increasing day length in the spring. Antlers are fully developed and ready for use during the autumn rut.

Antler size, as already mentioned, is affected not only by genetics but also by age and nutrition. Antlers can grow by 2 cm (0.8") per day, so the metabolic demands on males in the spring and summer are huge. Males have to browse and graze on high-quality food and need access to lots of minerals (such as calcium) to grow bone. As Darwin surmised, antlers are primarily used as ornaments, advertising the size and fitness of a male. Antlers are seldom used for defense or fighting – for that purpose, deer usually use their hooves and canine teeth.

Horns are very different from antlers. Horns have a bony core but are covered by a durable sheath made of keratin protein – the same stuff that makes our hair and nails. Unlike antlers, horns are usually unbranched, are not shed annually but grow larger throughout an animal's life, and often occur on both males and females (although those on males are usually larger). Horns occur on bovids (African buffalo and antelope such as gazelles, bison, muskoxen, and cattle), ovines (sheep, including bighorn sheep), and antilocaprids (North American pronghorn antelope). The largest naturally occurring horns known were found on extinct, long-horned bison (*Bison latifrons*) in Pleistocene North America at more than 2 m (6.6') long. Today, Asian guar

(*Bos gaurus*) have the largest horns at 1.2 m (4'), but Texas longhorn steers have been selectively bred to have horns more than 3 m (9.8') long from tip to tip. It proves that, like antlers, horn size is at least partly determined by genes.

Functionally, horns can also act as ornaments, but they differ from antlers because they are often used as weapons of offense or defense. For example, battles between Rocky Mountain bighorn sheep (*Ovis canadensis*), where males smash their coiled horns together, are legendary. Winners in these battles establish dominance and control over a harem of females, which leads to reproductive success. In many bovids, horns are used in battles of strength to determine dominance while minimizing injury. Although there are many differences between antlers and horns, both are important as secondary sexual characters and confer a reproductive advantage to males with the largest, most symmetrical ornaments.

Other Ornaments

Besides antlers and horns, other mammal ornaments include teeth that form the tusks of elephants (elongated incisors), walruses (upper canines), male narwhals (single left canine tooth), beaked whales, and pigs (two pairs of canines). As usual, genetics, age, and nutrition affect the size and symmetry of tusks. For instance, the effect of genetics is shown by elephant tusks, which are getting progressively smaller due to unnatural selection – ivory poachers are selecting against long-tusk genes by killing the biggest animals they can find and thus removing them from the gene pool.

What Messages Do Ornaments Send?

Male mammals with antlers, horns, and tusks use them to establish social order and dominance in order to establish territories, gain choice resources, and compete for mates. But what is the role of females? Are females simply passive, disinterested spectators, or are they affecting sexual selection? For most mammals, females have a lot more invested in reproduction than do males. During gestation, females carry and nourish young, which exacts nutritional demands. In addition, the birthing process can be dangerous and expose mothers and newborns to predators and infections, while after their birth, babies have to be weaned. Usually for male mammals, there is little invested in reproduction after conception. Because of this biological inequality, there is a lot of selection pressure for females to try to optimize their successful reproduction by mating with the fittest males (those that will sire the fittest, healthiest offspring).

Potentially, females may be getting lots of different information by assessing male ornaments. For example, if males are defending territories, then ornaments may reflect the quality of the territory – the abundance of resources there. Since ornament growth is affected by nutrition, males with larger ornaments are demonstrating that they have been in prime habitat. Thus, offspring will be born into a wealthy neighborhood. Ornaments may also reflect the ability of a mate to defend females and their offspring from predators or from dangerous members of their own species. Also, if it is older males that have larger ornaments, perhaps these experienced animals will be better at any kind of parental care males might participate in. Therefore, choosing mates with better ornaments should be tied to better survival of offspring.

Do Parasites Matter?

Do mammals with the best ornaments have the fewest parasites, and do parasites negatively impact ornaments? In an experimental study to try to answer some of these questions, female reindeer (domesticated caribou) with antlers were naturally infected with different kinds of parasites (including bot flies and gastrointestinal roundworms).[2] Some were then treated at the start of antler growing season with a broad-spectrum, anti-parasite medicine, while others were treated with a placebo as a control. At the end of the growing season, there was no association between parasites and length of antlers, but the females who received the medicine displayed increased antler symmetry – a measure of antler quality.

The relationship between parasites and ornaments was also investigated in African buffalo (*Syncerus caffer*). African buffalo are large, sub-Saharan bovids with bulls weighing up to 900 kg (1984 lbs) and sporting thick, lyre-shaped horns that can be more than 1 m (3.3') long. Cows also have horns, but these are shorter. Males use their horns in fights, which can be fatal, but death is rare. Instead, the horns are mostly used to display dominance, which gives higher ranking males access to females who are in estrus. African buffalo are polygynous – dominant males mate with several females.

In a 2008 study, parasite ecologists Vanessa Ezenwa and Anna Jolles found that horn size of male buffalo was a predictor of body condition; this trend was also found for females. About 87 percent of bulls and cows were infected with intestinal parasites, including roundworms and protozoan parasites called coccideans. For both males and females, buffalo with smaller horns had more species of parasites, and males with large horns had fewer kinds. In addition, bulls with larger horns had smaller populations of parasites. For females, horn size was also correlated with reproductive status – large-horned cows were more likely lactating. Female horns were an indicator that cows had successfully calved.

Overall, horns of African buffalo are honest indicators of body condition and parasite loads. Bulls can assess other males, getting information about their body condition and parasite infections, and they can also rate females. Cows with large horns are in better physical condition and can outcompete other cows. This information is important for males because a bull often has to shepherd a cow that is in estrus for several days to prevent her from being mated by another bull, thus foregoing other reproductive opportunities. For buffalo, where males mate with several females, females are usually the "choosy" sex. Dominant, healthy males aggregate with herds of high-quality (longer horned) females. In both sexes, horns are signals of parasite infections, health, and reproductive ability. From an evolutionary viewpoint, this information is invaluable because it is honest – animals are not getting fake news about parasites and the fitness of other animals.

Other Examples

Male Asian elephant (*Elephas maximus*) tusks grow throughout life (especially in the first 30 years), while females usually do not have tusks. In southern India, wildlife biologists found that males of the same age with longer tusks had fewer parasites. Tusks in elephants serve several functions. They can be used for feeding (removing tree bark), for predator defense, during agonistic encounters between males, and as sexual ornaments. Even for African elephants (*Loxodonta africana*), where both males and females have tusks, males have larger tusks and use them for display.

Female African elephants choose to breed with large, older, longer tusked males that successfully win male-male competitions. In Asian elephants, tusk size was less important to females than body size or "musth" (a temporary physiologically heightened sexual state in males that win competitions). Thus, tusk size (although an indicator of parasites) may be less important as a signal to a female Asian elephant than clues such as a male's physiological state. It seems that the use of ornaments to advertise parasites is less important than other features in some mammals.

In fact, unlike the prediction of Hamilton and Zuk, ornaments and parasites may be positively related, where animals with more elaborate ornaments have more parasites. For instance, in the mountains of Spain, wild goats called ibexes (*Capra pyrenaica hispanica*) occur. Males have horns that take about nine years to grow to a large size (about 60 cm or 23.6"). During the rut, when males killed by hunters were examined for parasites, it was found that goats with the best horn index (a measure of length, girth, and symmetry) were shedding the greatest number of lungworm nematode parasites in their feces.[3] The trend was not found for other parasites (other roundworms, tapeworms, or protozoans).

The unexpected trend of goats with better horns shedding more parasites is likely due to the special life cycle that lungworms have. Adult worms live in the main passageways (bronchi and bronchioles), where they produce eggs that hatch; juvenile worms then crawl up to the trachea, induce coughing, are swallowed, and eventually are passed out from the digestive system. However, sometimes juvenile worms will penetrate into lung tissue and remain dormant there until their host's immune system gets depressed. The worms then become activated, break out of the lungs, and are passed from the host. When goats are in the rut, this activity greatly increases stress on males, especially larger horned, dominant, 9- to 15-year-old adults. Increased stress causes the animals to release more stress hormones (corticosteroids), and males also have lots of the sex hormone testosterone. Both of these hormones cause a reduction in immunity, especially in dominant males. As a result, goats with the best horns are under a lot of stress and end up releasing the most parasites – the opposite of what Hamilton and Zuk would predict. No one knows if parasites acquired earlier in life (before nine years) might affect horn growth, but a study by Luzón et al.[4] shows that the relationship between parasites and ornaments is much more complex than we would imagine since different kinds of parasites may act differently from others.

Honesty in Advertising

If parasites do impact mammal ornaments, and if ornaments confer information about parasites, then the type of parasite is likely important. Ideal parasites to show the trend would not be too damaging to the host (very sick hosts would be obvious to other animals) and would likely cause persistent infections. Because hosts are continuously adapting to control parasites, and parasites must evolve measures to continue to infect hosts, the coevolutionary "arms race" leads to selective mating for resistant mates. Parasitic worms that infect various internal organs (for example, thorny-headed worms, tapeworms, flukes, and roundworms) and have life cycle stages outside a host, but must be transferred to naive hosts, will evolve to infect the most common genetic type of host they encounter – an average host. These kinds of parasites should be the kind that would best result in a sexual selection effect, where ornaments do honestly advertise parasite-infected hosts. In these cases, males cannot hide the fact that they are "wormy" and that they may have poorer disease resistance than other potential mates.

The Fix Is In: Parasites and Male-Male Competitions

Besides parasites affecting ornaments that can be read by choosy females, another complicating factor occurs if parasites affect the outcome of male-male competitions. Often during the rut, male deer establish breeding hierarchies. Body condition, body size, stamina, and antler quality all play a part to determine a particular male's place on the mating ladder and, therefore, his breeding opportunities. The metabolic demands and stress on males during the rut can be so great that body condition and weight are reduced – body fat composition can reach starvation levels. What happens when males are also infected by parasites? Do parasites impact the placement of males on the breeding hierarchy?

To see if parasites could affect male competitions, parasite ecologists studied white-tailed deer in South Carolina.[5] These deer were infected by giant liver flukes (*Fascioloides magna*; see chapter 4), and by age four, 66 percent of all deer were infected. The worms live in thin-walled capsules in the livers of deer; adult flukes have little obvious disease effects on white-tailed deer, but the infections last for months to years. As a result, liver flukes are just the kind of parasite that ticks a lot of the boxes for those that should be related to ornaments.

When interactions among stags were examined, fluke-infected male deer were about 5 percent lighter in body weight and had significantly fewer antler points than uninfected animals. The effects of flukes on these characteristics were even more pronounced when the intensity of infection (the average number of worms in infected hosts) was taken into consideration. This study suggested that parasites were having a negative effect on characteristics that are crucial in determining mate competitions and the breeding hierarchy in deer.

All Things Considered

Taken together, studies show there are links between parasites, ornaments, immunity, and reproductive behavior, but like most things in biology, the links are not simple or direct and do not easily allow for predictions. One study that set out to harmonize all these factors looked at parasites of Grant's gazelle (*Nanger granti*) in Kenya.[6] Grant's gazelle are grazers that occur in the grassy plains and shrublands of eastern Africa. Both males and females of these beautiful gazelles have gently curved, ringed horns, reaching lengths of 80 cm (31.5") in males. Grant's gazelles are polygynous – males defend territories that determine access to females and reproductive success.

Males who defend territories have more gastrointestinal roundworms (more diverse species and larger populations) than males that do not. This trend could be because the parasites are directly transmitted – gazelles consume infective stages on pasture, so defending and staying on a territory would result over time in a locally high concentration of infective stages.

However, the trend could also be due to the physiological stress that territory holders are under since stress has a negative effect on immunity. Perhaps both factors are at play – territory holders are exposed to more parasites, and their immunity is weakened.

In male Grant's gazelle, the level of testosterone was directly related to horn size and to the frequency of territoriality – males defending territories had larger horns and more testosterone. In addition, testosterone levels were positively associated with levels of stress hormones – glucocorticoids. So, how do these factors – sex, immunity, parasites, and ornaments – come together?

The effect of testosterone on males seems to be a "double-edged sword." More testosterone equates to better reproductive success, but it can depress immunity and lead to more parasites. However, opposite to what we would guess, testosterone in male gazelles was positively related to immunity – but to a particular kind of immunity. Innate immunity includes defenses that act immediately and include various nonspecific resistance mechanisms that are encountered by parasites the first time they infect a host. As the level of testosterone increased, animals had better innate defenses. However, adaptive immunity is the acquired specific defense that kicks in after a first infection, involves special white blood cells, and takes longer to develop. It had the expected relationship with testosterone – as hormone level increased, immunity strength was reduced. Consequently, effects of testosterone on immunity are not always suppressive and, in some circumstances, may even strengthen defenses against parasites.

Overall, therefore, male secondary sexual characters (ornaments) seem to be positively affected by testosterone and result in more reproduction, but because the effect of testosterone on immunity varies, males with better ornaments likely have smaller infections of some parasites and better innate resistance but larger infections of other kinds of parasites. Females that preferentially mate with highly ornamented males are likely getting mates that will pass down genes resulting in healthier offspring. But it seems that the onus is on female mammals to assimilate a lot of information in order to choose the best mate. As a result of having to process so much information, some biologists have suggested that female mammals develop more intelligence than males – that seems to be the case for female Australian bowerbirds, which assess males based on courtship displays, decoration of bowers, and parasites.

So What?

In 1871, I am sure that Darwin never dreamed that ornaments, hormones, behavior, stress, and parasites could all be so intricately entwined in explaining features like the antlers of Irish elk. To discover if mammalian ornaments are honest indicators of parasites and health, each kind of mammal, and the kind of ornament they use (antlers, horns, tusks), matters. Also, the specific kinds of parasites, how much damage they cause, and how they are transmitted to other hosts all need to be considered. Generally, people don't like answers that are complicated and messy – they want simple black-and-white explanations. Science, however, usually discovers that simple answers do not adequately describe what happens in nature. When Hamilton and Zuk suggested that parasites can have a significant effect on sexual selection, they opened a large can of complicated worms. Nevertheless, although we cannot simply conclude that mammals with better ornaments have fewer parasites and diseases, their idea was very valuable because it stimulated many studies of wild mammals and their parasites, and these have resulted in great gains in our knowledge of the natural world and how it works.

9
The Night of the Vampire: Parasitic Mammals and Bat Bugs

It seems that most people are preconditioned to loathe parasites and to find them disgusting, dirty, and gross. In classes when I was teaching, just saying the word "tapeworm" or "lice" usually caused looks of disgust and consternation among my students (unless this look was actually a comment on my teaching!). Couple this reaction with the innate fear that many have for some animals – like spiders and snakes – and I guess it is not surprising that vampire bats would be particularly feared. Bats, including vampire bats, are mammals; yet few mammals have become parasites, in spite of it being so common a biological lifestyle among other classes of animals. How (and why), then, did vampire bats become blood-feeding parasites? Well, it turns out that other parasites (in this case blood-feeding insects) could have been the reason.

The Vampire Legend

In the 1890s, an Irish theater manager named Bram Stoker met a Hungarian traveler, Ármin Vámbéry, in London. Vámbéry regaled Stoker with stories from the Carpathian Mountains, and in 1897, Stoker wrote the gothic horror novel *Dracula*. Stoker was likely influenced by another Irish horror novelist, Joseph Sheridan Le Fanu, who 26 years earlier had published a novel about a female vampire, *Carmilla*. Vampires were undead beings who came back from the grave to steal the souls of their victims by drinking their blood.

The stories may have arisen because of a common disease in Eastern Europe called consumption, likely tuberculosis, which is caused by the bacterium *Mycobacterium tuberculosis*. These germs have wax in their cell walls, which is hard to make and causes the bacteria to have very slow reproduction, leading to a lengthy disease. Family members watched helplessly as a loved one developed a hacking cough, leading to a prolonged illness in which the person wasted away. After the funeral, someone else in the family would start coughing, invariably leading to the same fate. Superstitious people thought that the dead relative had come back to steal a soul, and this belief may have contributed to legends of monsters who come in the night to take blood. Another possibility is that both consumption and the vampire myth were due to rabies.

Real Vampires

Because the parasitic lifestyle is so pervasive in biology, it should not be a surprise that even some mammals have taken it up. Regardless of where and how the legends arose, there are three decidedly real species of blood-feeding vampire bats alive today. Two species are rare and feed mostly on birds – *Diaemus youngi,* the white-winged vampire, and *Diphylla ecaudata,* the hairy-legged vampire. The third species is the "common vampire," *Desmodus rotundus,* which feeds mainly on mammals. Another three species of extinct vampire bats are known from fossils. However, bats cannot be the source of the legend since they were given their common name due to their resemblance to the monsters, not the other way around.

Today, vampires occur in warm tropical and subtropical habitats stretching from Mexico to Argentina. Extinct species once occurred as far north as the southern United States, from California to Florida. Common vampire bats occur in a variety of habitats from coastal areas to forests, but they do best in agricultural areas with lots of domestic animals to feed upon. Safe diurnal roosting sites, like hollow trees, caves, and abandoned buildings, determine bat abundance. Good roosting sites are crucial for vampire bats because their roosts are where social interactions, reproduction, parental care, food digestion, and escape from predators or inclement weather occur.

Bats are perplexing mammals for evolutionary biologists and paleontologists to understand. Due to their small size, fragile skeletons, and tropical habitat, the fossil record of bats is spotty. They appear in the evolutionary record fully developed for flight some 65 million years ago. They are related to shrews and likely evolved flight from the forest canopy to the ground, perhaps first as gliding mammals. Specifically, vampire bats are related to leaf-nosed, mostly insect-eating bats. When and how bats took up a blood-feeding role is puzzling – there are at least four different ideas about how bats became parasites.

One idea is that vampire bats evolved from ancestors that fed on blood-feeding external parasites. If true, it's possible that their first taste of blood may have been their own. Social bats do a lot of allogrooming – picking off and eating ectoparasites like ticks and insects called bat bugs (similar to our bed bugs) from each other. If a blood-engorged insect or tick was consumed, the groomer would get an extra dollop of protein along with its arthropod meal.

A second hypothesis is that vampire ancestors fed on insects that were attracted to weeping wounds on large mammals. Large mammals often have wounds from failed predator attacks, clashes with members of their own species, and accidents. Insect-eating, pre-vampire bats would have been attracted to the swarms of insects congregating at the wounds but could have switched to feeding directly on the wounded mammal's bleeding tissues.

A third suggestion is that vampire bats evolved from fruit-eating ones, which have sharp, well-developed incisors used for cutting into the tough rinds of tropical fruits. These incisors would later become the razor-sharp incisors of vampires. This idea, however, provides no explanation of why vampire ancestors would have changed their diet from fruit to blood.

A fourth theory proposes that vampire bats evolved from bats that were predators of small animals (such as birds, bats, rodents, and lizards) up in the canopy of the rainforest. As support for this idea, today two of the three species of vampire bats primarily feed at night on birds that are perched in trees.

Based on the fossil record, some leaf-nosed Neotropical bats were carnivores living about 10 million years ago, just when vampires presumably first evolved. Some evidence suggests that this period was a time of climate change, when South America was becoming more arid and large continuous tracts of forest were being broken into habitat islands. These forest clumps would have been havens for marsupials, primates, sloths, other mammals, and birds, which would have become concentrated and easier for predatory bats to find. Many of the concentrated animals are too large for a bat to predate upon, but bite wounds would have provided nourishment. Animals sleeping in trees that could be wounded with little pain would be unaware of being attacked and would remain still long enough to provide a free-flowing stream of blood. This situation might have selected for many of the adaptations that are seen in vampire bats today.

Vampire Adaptations

However the parasitic habit evolved in bats, like most parasites, vampires have many specializations. Today, all are sanguivores – they must have blood meals to survive and reproduce. Blood is an interesting and unusual food containing about 80 percent water and 20 percent protein – there is virtually no fat and very little carbohydrate. As a result, bats cannot store energy as fat and must have a new blood meal every couple of days. Vampire bats cannot fly long distances searching for hosts (that would burn up too much energy), so their food supply must be near, abundant, and predictable.

To feed, vampires have very sharp modified incisor teeth that make a divot-like wound in a host, causing little pain. Like most blood-feeding parasites, the saliva of vampires includes proteins that prevent blood clots and stop torn blood vessels from constricting.

Other adaptations that vampire bats have are a long and extendable grooved tongue, notched lips, and a gap between their incisors to help them drink their liquid diet. Consumed blood goes into a short esophagus, ending in a forked tube that can cause the blood meal to bypass their stomach. Mammalian stomachs evolved as food-storage organs but also serve another important function: stomach acid helps to break down food but, more importantly, helps to kill most ingested dangerous bacteria and parasites. However, blood is liquid and heavy, and bats have no need to store it, so vampire bat stomachs serve a special function – they allow for rapid fluid absorption and elimination of heavy water so the bats can fly again soon after a blood meal. However, the specialized stomachs may not kill as many microbes as do other mammal stomachs.

The water from a blood meal goes directly to the kidneys, and because water is heavy and inhibits flight, vampires begin urinating within a few minutes of feeding, often while they are still on their host. This process is so efficient that about one-quarter of a blood meal is excreted as urine one hour after feeding. Vampires also lose respiratory water when flying, so they must live in humid areas – they would rapidly become dehydrated if they lived in desert habitats.

Natural History: The Night of the Vampire

Common vampire bats prefer to feed during the darkest, least moonlit times. Using echolocation and smell, they usually approach their sleeping victims from the ground and have modified legs and wings that let them scrabble quickly to their host. They hop onto the victim and use specialized heat detectors in their noses to identify blood vessels that are close to the skin surface. *Desmodus* uses its cheek teeth, like scissors, to clip away fur and then makes a painless bite with its incisors. Humans sleeping in hammocks are often bitten painlessly on the feet or toes without waking. Blood pools from the wound and is then lapped up. After feeding, vampire bats launch themselves off the host using their extra-long thumbs. They may temporarily roost not far from their host and eliminate urine to reduce their weight before flying back to their daily roost.

Early in their evolution, common vampires probably exploited large ground sloths and cow-sized armadillos (glyptodonts), and before the introduction of domestic animals such as docile horses, cows, and pigs, *Desmodus* fed mostly on large ground-dwelling mammals like white-tailed deer, tapirs, capybaras, and peccaries. But now, with the advent of agriculture and domestication, vampire bats are blessed with a constant and predictable source of blood, and their population has expanded greatly.

Common vampire bats are very social, roosting in socially segregated colonies that can include hundreds of bats. Females in harems, and their offspring, roost near a

Figure 9. An adult bat bug (*Cimex* sp.). Source: matuty/stock.adobe.com.

dominant male. Satellite males form small groups nearby, hoping to sneak in a copulation when the dominant male isn't vigilant. Males will fight using their canine teeth for access to female roosting areas. Because of their high metabolic rates and inability to store fat, vampires must feed often, and it has been reported that, when they return to their roost after feeding, they sometimes regurgitate and share food with other bats, especially relatives, thus reinforcing social bonds. Also strengthening social groups, vampires spend about 5 percent of their time grooming themselves and other bats, again preferring relatives for this activity. It turns out that, for vampire bats, grooming is a very important activity because they themselves are fed upon by their own suite of roost-dwelling, blood-drinking ectoparasites – bat bugs (Fig. 9).

Parasites of Parasites

Vampire bat bugs are an example of "hyperparasites," parasites that feed on other parasites. Although we have nightmarish legends about undead blood-feeding vampires, the highly specialized vampire bats of the tropics have their own, very real nightmares to deal with.

Bat bugs are flat, oval, small, and wingless, resembling apple seeds. Like the vampires that are their hosts, bat bugs are "obligate" parasites, meaning they only feed on blood. You may be familiar with the type that feeds on us – bed bugs – but colonial birds (especially swallows) and mammals that live in humid, dark, enclosed spaces, like vampire bats, are also attacked by specific species.

Adult bat bugs live in cracks and crevices in the roosts and feed repeatedly on a host.

They lay eggs off the host, which hatch to produce several generations of nymphs, all of which take blood meals before they molt to the next life cycle stage. As a result, a single bat bug may take many blood meals during its life. Adult bat bugs are very patient and persistent, and can live for months without food, biding their time until their hosts return.

Populations of bat bugs can build up to amazing numbers in roosts that are used continuously. As people who have been bitten by bed bugs know, the bites are very irritating and can result in enough blood loss to cause anemia. Bat bugs can also be important for spreading microbial diseases. For example, Bartonellae are bacteria that infect the cells that line blood vessels and red blood cells of mammals, including bats (one disease they cause in humans is trench fever). More than half of the common vampire bats examined from the rainforest of Peru were found to be infected with *Bartonella*. The bacteria also occur in bat bugs, and because vampires feed on mammals, the infection can be spread from bugs to bats and then to other mammals, including humans.

Bat bugs may be an important selection pressure that has led to enhanced social behaviors, including allogrooming in vampire bats. The bugs can also affect the ecology of the bats – for instance, when populations of bat bugs get exceedingly high, they can force bat colonies to abandon roosts.

Why Aren't There More Vampires?

Based on how pervasive parasitic lifestyles are in the animal kingdom, it may seem strange that, with their success and diversification, even more mammals have not taken up a blood-feeding parasitic existence. The key to this conundrum is that a diet composed only of blood is very challenging. Other than water, blood consists of protein (93 percent) and little else. Vitamins, minerals, lipids, and other nutrients are rare or missing, so having a balanced diet is an issue. Also, blood-borne pathogens commonly occur, especially viruses that can be deadly. Therefore, with their specialized diet that is nutritionally incomplete and the fact that their only source of nourishment can result in infectious diseases, vampire bats have evolved three unique adaptations for dealing with these problems.

First, the genetic material of vampire bats is unique. It includes lots of small pieces of DNA called transposons or "jumping genes." These genes can move around in the chromosomes of vampire bats and affect the expression of other genes – turning some genes on, others off, and even causing new functions. As a result, vampire bats have few genes involved in bitter and sweet tastes (which are important in fruit-eating bats), and vampires have special genes that affect how they balance blood sugars.

Second, researchers have discovered that other jumping genes in vampires are related to immunity and defense against viruses and also to lipid and vitamin metabolism. In addition, the vampire bat genome has many DNA sequences that have come from viruses inserted into it. Surprisingly, however, these virus DNA insertions are fewer than the number found in non-blood-feeding bats – even though vampires are exposed more often to dangerous viruses. Somehow, in a way that we do not yet understand, vampire bats have developed adaptations that prevent the proliferation of virus genes within their own genomes. The result is that the vampire genome does a good job of protecting bats from viruses – for instance, they have more antiretroviral genes (genes to prevent infection by viruses like HIV) than do any other bats. The relationship between bats and viruses is complex and poorly understood, but based on our current COVID-19 pandemic with coronavirus, which may have originated in bats (but not vampire bats), we must do more research to prevent future pandemics and, perhaps, to find out about bat adaptations for handling viruses.

A third unique adaptation of vampire bats is the special community of microbes that live in their intestinal tracts (the microbiota). These bacteria help bats with blood meal digestion and also supply necessary vitamins that are missing in the vampire diet. In addition, the bacteria enhance vampire immunity. The microbiota (also called the normal bacterial flora) of all mammals is affected by diet and evolutionary history, and the bacterial community found in vampires, although most similar to that in insect-eating bats rather than in carnivores or fruit eaters, is completely distinct and unique. There is very little variation in it among different individual vampire bats – every vampire bat in a colony has the same normal flora. This observation sheds some light on the social biology of vampire bats. The essential role the normal flora bacteria play shows that social regurgitation of food is critical to ensure that young bats can establish their own microbiota and thus be able to feed on blood.

In taking up a blood-feeding life, vampire bats have not only had to evolve many structural, genetic, and behavioral specializations but have also had to coevolve with a unique suite of microorganisms and contend with their own hyperparasitic bat bugs. Their hyperparasites may even have played a crucial role in the evolution of vampire bats by encouraging social behaviors that enhanced colonial life and promoted regurgitative food sharing. These behaviors could in turn have contributed to the development of the unique vampire microbiome and assisted in selecting genes to control pathogenic viruses. As weird as it seems, perhaps the reason there are not more blood-feeding mammals is because of the unique life and host specializations between vampire bats and vampire bat bugs. When we consider all the problems and the many requirements and specializations that being a blood-feeding parasite involves – including the apparent need for an equally specialized hyperparasite – it makes sense that more mammals have not evolved to exploit this lifestyle.

Earning Respect

As I write this book, no one knows the origin in humans of the coronavirus, which has caused our COVID-19 pandemic resulting in millions of deaths, but one suggestion is that it came to us from bats. Bats host many pathogens that can spread to humans and other animals – bacteria, protozoans, fungi, and viruses. The zoonotic infection agents bats carry include more than 50 dangerous RNA viruses, like rabies, SARS (severe acute respiratory syndrome), and Marburg virus, which causes a hemorrhagic fever with more than 80 percent mortality. Bats themselves are suffering from an outbreak of a fungal disease, white nose syndrome, that threatens their populations in North America, but they have evolved amazing mechanisms to suppress many of the viruses they carry. It is not surprising that many people hate bats, but perhaps if we let them live in their natural habitats, and learn more about their biology, we will be able to discover how they deal successfully with viruses. We may then be able to turn some of this knowledge into treatments and cures.

Because vampire bats affect animal husbandry and can transmit infections, they have contributed to the fear and loathing that many people have for bats in general. This attitude has led people to poison bats, destroy roosts using explosives and napalm, and declare war on these remarkable little mammals. However, when we consider how many specializations in body structure, behaviors, genomes, and ecology that vampire bats have had to evolve, we should have at least a grudging respect for them. The challenges of being a blood-feeding parasitic mammal are daunting. Couple that with the need for a reliable host supply, finding suitable roosts, and dealing with ecosystems that are rapidly changing, and I think that, rather than fearing them, we should appreciate vampires, marvel at how this evolutionary experiment began, and wonder if it will be able to continue.

10
Your Brain on Worms: Nature's Biological Weapon

Parasites use hosts for several purposes – as a source of nutrients, as cozy homes, and as ways to get their progeny into more hosts – but examples of ways that hosts can turn the tables and use parasites to promote their own survival are not as common. One way that hosts can take advantage of parasites, especially those that have the ability to cause lots of damage, is to turn them against their ecological competitors – as biological weapons – as in the case with brain worms of white-tailed deer, which can kill other animals, even though they don't cause many problems for the deer. In addition, our continuing interference in nature through anthropogenic climate change and destructive extraction of resources is facilitating the spread of these worms and making us accessories to the death of moose, caribou, muskoxen, and other species.

Moose Sickness

As a helicopter flew over mixed deciduous-coniferous forest in northern Ontario, surveying moose to determine the sustainable harvest for hunters, the pilot was the first to see it. Over his radio, he pointed out to the wildlife biologist on board that an adult bull moose was in a boggy clearing below. Tracks in the snow clearly showed that the moose had been repeatedly walking in a circle. Its neck was held at a crooked angle, one of its ears was tilted up, and it was staggering like a drunk. "Another case of brain worm," the biologist explained. "Soon the moose will collapse and die, if wolves don't get him first."

Moose sickness (cerebrospinal nematodiasis) is caused by a roundworm (nematode) parasite with an unwieldy name, *Parelaphostrongylus tenuis*. It belongs to a group of worms called Protostrongylids. Most of these parasites infect the lungs of their hosts, especially deer, but not *P. tenuis* – it lives in brains.

The life cycle of brain worms involves terrestrial snails and slugs as intermediate hosts. Small first-stage juvenile worms, which have a characteristic spine on their back surface, are passed in deer feces (Fig. 10). On the forest floor, these juveniles penetrate the foot muscles of snails and slugs that are attracted to deer feces. Once inside snail foot muscles, the worms molt twice and grow to become slightly larger stages – the third-stage juveniles. This stage is the only one that can infect deer. When an infected snail or slug is accidentally eaten by a deer, worms released by digestion in the gastrointestinal tract penetrate through the gut lining

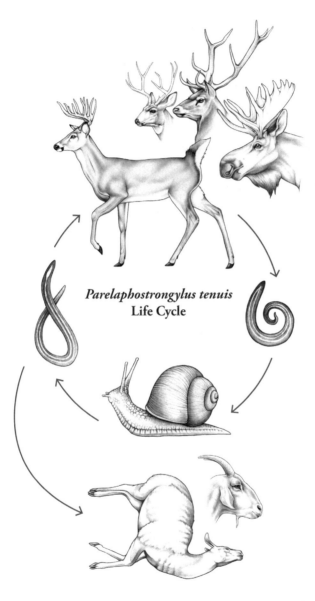

Figure 10. Life cycle of *Parelaphostrongylus tenuis*. Source: Illustration by Henry Crawford Adams, Lincoln Park Zoo.

to enter the host's body cavity. After migrating to the spinal nerves in the deer's lower back area, worms enter the spinal cord. It takes about 10 days from when snails were eaten until the worms appear in the spinal cord. Now in the central nervous system (CNS), worms molt twice to sub-adult stages and crawl forward toward the cranium. Eventually *P. tenuis* reaches cranial veins and venous spaces between the meninges (connective tissue lining of the CNS). Finally home, brain worms mature to adult male and female worms.

When both mature male and female worms occur in a host's brain, unembryonated eggs will be carried in blood vessels, first to the deer's heart and then to its lungs. First-stage juveniles develop, hatch, and are swept up the trachea of the lungs with mucus. Once swallowed, they pass through the entire intestinal tract and are passed with the host's feces. The little first-stage juveniles, with their characteristic dorsal spines, are the only life cycle stage that occurs outside a host – they must survive environmental conditions on the forest floor until they can get inside a snail.

Relatives

Besides brain worms (also known as meningeal worms), two other closely related species occur in North America with unusual distributions, different hosts, and living in different sites in their hosts. Unlike *P. tenuis*, which occur in the meninges of the central nervous system, the other species live in skeletal muscles. Identifying infected hosts of all three types of these nematodes, and discovering their geographic distributions, has been very difficult. Just looking at mammal fecal samples for dorsal-spined juveniles is misleading because hosts may be infected by one or more kinds of Protostrongylids (all of which have the same dorsal spines at that stage). Also adding to the confusion, a host can be infected with adult parasites but may not be passing any juveniles in its feces (resulting in what are called "occult infections").

Performing postmortem exams of mammals for adult worms involves dissecting hosts and finding male worms. Unfortunately, only adult male roundworms have the definitive features that are needed to conclusively identify the worms to species. As you can imagine, this process is really challenging for finding muscle worms because there is so much tissue that needs to be processed for each host, and the worms are so small (the parasitological equivalent of finding a needle in a haystack!). As a grad student, I well remember sharing a lab with very dedicated, hard-working wildlife parasitologists, who painstakingly picked through piles of deer meat using lighted magnifying lamps in order to find worms. Luckily, today molecular methods of species diagnosis makes the task easier, faster, and probably more accurate.

Using these methods, researchers discovered that white-tailed deer (*Odocoileus virginianus*) in the southeastern United States were infected with a muscle worm named *Parelaphostrongylus andersoni*.[1] There is now evidence that it occurs widely throughout North America, wherever white-tailed deer occur, and there is also a population that infects barren ground caribou (*Rangifer tarandus*) from Alaska to Newfoundland.[2] The occurrence of *P. andersoni* muscle worms in caribou likely represents a host switch event that occurred during the Pleistocene – worms successfully colonized a novel host.

In spite of the great overlap in distributions of *P. andersoni* muscle worms and *P. tenuis* brain worms in eastern North America, wildlife disease experts have shown that the two worms are ecologically segregated – when deer are infected with *P. tenuis*, they are never infected with *P. andersoni*. Perhaps this segregation could be due to the forest floor microclimate having different effects on the survival of young worms of each species, but a more likely reason is cross-immunity, where infection by one worm means the host develops an immune response against subsequent infection by the other worm.

A number of other hosts, such as mule deer, Columbian black-tailed deer, mountain goats, and woodland caribou – but not white-tailed deer – can become infected with another muscle worm, *Parelaphostrongylus odocoilei*. This parasite overlaps with *P. andersoni* and occurs in western North America as far east as west-central Alberta. This muscle worm has become specialized to infect black-tailed and mule deer in the west, and it causes little damage to its hosts. To summarize, although there are three closely related roundworms that infect muscles and brains of deer, environmental features, host specificity, and immunity all keep them ecologically separated. By the way, hunters can have no fear of eating venison with muscle worms – you cannot be infected – but for grazing mammals that ingest snails and slugs, the consequences can be devastating.

Consequences

Brain worm infection is common in white-tailed deer in the eastern and midwestern parts of the United States and Canada, and it has also been found in Costa Rica, perhaps due to a southern range expansion by infected hosts. Notably, there appears to be an invisible barrier that prevents brain worms from occurring in the west-central parts of North America and in the far north, maybe because these areas are too cold for survival of early life stages – thus preventing transmission. Perhaps host immunity to the species of Protostrongylid worms that routinely occur in the west inhibits brain worms and has stopped them from a westward spread. Whatever the reason, it is certain that, if brain worms ever did successfully colonize the west, the ecological and environmental consequences would be disastrous because brain worms, although mostly harmless in white-tailed deer, cause moose sickness and kill other native cervids as well as bovids, bighorn and other sheep, mountain goats, livestock, and even exotic animals like alpacas and llamas.

In the brains and spinal cords of grazers and browsers other than white-tailed deer, *P. tenuis* worms cause inflammation and degeneration of nervous tissues. Animals with moose sickness sway when they walk, have weak, unsteady hindquarters, tilt their head and necks to one side, go in circles, and become fearless and blind. Sick animals have impaired muscle movements and suffer from partial paralysis and weight loss. Males will sometimes have abnormal antler development. Susceptibility and resistance to the disease varies among species – moose and elk seem to cope better with the infection than bighorn sheep or mountain goats, but even for them, the outcome of infection is usually not good. Caribou are very susceptible and at high risk – the infection in them is so dangerous that caribou cannot be reintroduced into habitats with a significant density of white-tailed deer without dying from brain worm.

What causes the differences in response to brain worms? We think that white-tailed deer and brain worms have a long, coevolutionary history. It is likely that white-tails mount a protective immune response to the worms, but the trick here is that the immunity is concomitant, which means that adult worms resident in the brains of deer stimulate immunity that stops new infections, and thus deer can't be infected with lots of worms. Although the proportion of deer in a population that are infected by brain worms can be as high as 91 percent of fawns by the time they are 18 months old, the average number of worms per deer is small. As a result, fawns and yearlings usually have two or three worms, and this number only climbs to four worms in white-tailed deer that are 7 to 15 years old. Also, if the worms in one deer all happen to be the same sex, then no eggs (and the resultant juveniles) can be produced. The small number of worms found in white-tailed deer do not cause much inflammation and central nervous system damage, and the worms can survive and reproduce for years.

In hosts other than white-tailed deer, however, worms in the central nervous system induce a strong host inflammatory response that causes lesions and damage. Not surprisingly, the frequency of moose disease is directly related to the density of white-tailed deer in an area. How much moose disease occurs in an area is also affected by climate – areas that get more precipitation and have more frost-free days have more moose disease.

A Secret Weapon

There has been an increase in the white-tailed deer population in North America and an expansion of their range from historical times. One factor usually attributed to explain this increase has been the persecution of wolves. As a top predator, wolves contributed to limiting the population of deer, but when we declared war on wolves and tried to eliminate them from most of North America, this limiting factor on the population of white-tailed deer was removed. In addition, white-tails do well living in areas where humans have modified the environment, and they often thrive in urban areas. As a result, white-tailed deer are expanding their range by about 1 km (0.6 miles) per year.

Another limiting factor for deer, however, is competition for food and other resources, particularly with moose, as large moose can exclude smaller deer from a habitat. But white-tailed deer may be able to use their brain worms as a secret biological weapon of competition to infect moose, other wild cervids, and even domesticated animals. If *P. tenuis* ever gained a foothold in western North America, many native species would be at great risk. Although populations of elk and moose seem to be currently stable, introduction of moose sickness might tip the scale against them.

Woodland caribou (*Rangifer tarandus*) in Alberta and British Columbia are especially vulnerable. Herds of woodland caribou are declining drastically, and today several herds are teetering on extinction. An important food for caribou is lichens, which can grow in sites with few resources (such as the surfaces of rocks) but are slow growing, so woodland caribou need large tracts of undisturbed, old-growth forests in order to have enough food to keep healthy, stable populations. Habitat destruction and fragmentation by humans for industrial activities (like oil and gas extraction and forestry) has reduced the usable habitat for caribou, while roads and cut lines have created easy access routes for predators, especially wolves.

Wildfires are also a problem because in burned areas, old-growth forest is replaced by younger vegetation, which is ideal food for deer and moose. As burned-out old-growth forests regrow, increasing populations of deer and moose attract wolves, which also kill caribou. Forest insect pest outbreaks like mountain pine beetles (likely due to a warmer climate) cause mature trees to die, and the deadfall in turn provides more fuel for fires. The number and intensity of forest fires are increasing each year, with a key driver being more frequent droughts caused by climate change, aided in western North America by outbreaks of pine beetles.

If climate change results in milder winters in western North America, it may cause parasites to also become players in this complex game, one with the potential to put the final nail in the coffin of woodland caribou. Currently, white-tailed deer in the west are not infected with brain worms, but if that changes and the prevalence of muscle worms is reduced, white-tailed deer invading caribou habitat will have a devastating impact, and brain worms will be the secret biological weapon that unleashes the destruction. Deer in other areas of North America are already killing moose and elk with brain worms.

Besides caribou, another iconic mammal of the north may be at great risk from a relative of brain worms: muskoxen (*Ovibos moschatus*). These are hardy, grazing animals that are culturally significant and also very important as a food source for many remote communities in the Arctic. Recently, the Canadian and Alaskan population has declined from about 1800 to 900, and hunters noticed that many animals were showing signs of stress – they had labored breathing, bled from the nose, and reduced ability to run for long distances. When wildlife disease experts looked for parasites and diseases, they discovered that the animals were infected by a type of lungworm, the protostrongylid roundworm *Umingmakstrongylus pallikuukensis* (this tongue-twister gets its inspiration from the Inuit language: *umingmak* – muskox; *pallik* – an arctic region; and *uuk* – a river).[3]

Unlike the brain and muscle worms we've seen, these parasites are similar to most protostrongylid roundworms in that they live in the lungs of their hosts. The lungs of muskoxen can have adult worms, eggs, juveniles, and fluid that causes lung damage and predisposes animals to secondary bacterial infections. Juvenile worms are passed with the feces of muskoxen and get eaten by terrestrial slugs, such as the common garden slug (*Deroceras laeve*), which in turn reinfect muskoxen when they eat slugs while grazing. Lungworm disease seems to be increasing in muskoxen.

Herds in some areas were 100 percent infected; one bull had more than 250 cysts in its lungs and was passing about 2000 juveniles per gram of feces.

The development and survival of the juvenile worms depends on temperature – they need at least 8.5°C (47°F) to survive. We know that human-caused climate change is severely affecting the Arctic with unprecedented increases in temperature, and it seems likely that climate change is a key factor in muskoxen having greater infections of lungworms. Sick animals have less ability to escape predators, making them more susceptible to grizzly bears and further contributing to their dwindling numbers. Sadly, human disturbances of our ecosystems are so large and happening so fast that natural protection by mechanisms like adaptation and evolution, which have shielded mammals from problems like moose disease or lung infections for thousands of years, will prove inadequate to address these new problems. Although there may be a few animals left in zoos, without a serious change of course for the global climate, it seems likely that our children will never get to enjoy the sight of caribou in the mountains or muskoxen on the tundra.

11

The Tale of the Tape: The World's Longest Parasite

One thing biology students soon learn is that there are few absolutes in this science – there seem to be exceptions to every rule. When teaching, I am cautious and hate using words like "never" and "always." However, biologists have noticed general trends and patterns, and put these down in statements usually called "rules." For instance, in ecology there is Bergmann's rule (named for German biologist Carl Bergmann) – the idea that birds and mammals inhabiting colder regions will be larger than their counterparts in warmer zones. Parasitologists have also noted some general trends, and one of these is called Harrison's rule – the idea that larger hosts have larger parasites. Prepare to be wowed as we investigate some evidence for this rule and meet some parasites along the way that are truly massive.

Tapeworms

Tapeworms, which we first met in the introduction, are parasitic flatworms (phylum platyhelminthes, class Eucestoda) of which we know about 6000 species. They are the most specialized of all animal parasites, and they all live as reproductive adults in the intestines of vertebrates. They have become specialized by losing or reducing most features that occur in free-living animals – for example, tapeworms have no appendages, no sensory structures such as eyes, no circulatory systems, and no digestive systems. My PhD supervisor very appropriately called them "gutless wonders."

Tapeworms have unique bodies called strobili (sing. strobilus) that are well adapted for a parasitic lifestyle. For example, tapeworm anterior ends form holdfasts called scolices that can include hard, chitinous hooks, muscular suckers, grooves, clamps, and combinations of these, specifically shaped to allow them to attach to the unique architecture of their particular host's intestinal lining. Behind the scolex is a very metabolically active area, the neck, that continuously generates serially repeated body segments called proglottids. As the newly made proglottids form, they are pushed backward, so the tail end of the worm is the oldest part. As new segments are produced and move backward, reproductive organs develop inside. Most tapeworms are hermaphrodites, with each body segment housing multiple male and female reproductive organs. As a result, these parasites are little more than nutrient-absorbing reproduction machines.

As hermaphrodites, tapeworms can self-fertilize but prefer to cross with other individuals of their own species that happen to be present in the same host. As proglottids move further back, they become filled with shelled eggs, inside of which are embryos. These egg-filled "gravid" proglottids fall off from the posterior-most region of the tapeworm, and eggs inside the disintegrating worm tissues are passed out with host feces. Each tapeworm can produce thousands of eggs per day.

To provide the energy and resources they need to make so many eggs, tapeworms absorb rich, semi-digested host food directly through their body surface, like a sponge. Their flattened, ribbon-like body shape is ideal to maximize the surface area they need for this absorption, and at the microscopic level, their tegument (their skin, which we came across in liver flukes in chapter 4, that acts as a host-parasite interface) is thrown into millions of small, finger-like projections, called microtriches, that greatly enhance its absorptive ability. Their major life challenges, once inside a mammal, are to hold on and not be forced out by strong muscular waves generated in the intestine; to not be digested by the host; and to fight against any immune attacks that their host may mount. In small infections, these quintessential parasites cause little harm in mammals – most live quiet lives, stealing trivial amounts of nutrients from their hosts. However, that's not always the case, as we shall soon see.

To get from inside one mammal into the intestines of another, tapeworms have evolved complicated life cycles that often use several kinds of intermediate hosts. Tapeworms are transmitted trophically, that is, they infiltrate the food web of their final vertebrate host and enter that host's body by being eaten. After being passed out of their mammalian host, tapeworm eggs are eaten by an invertebrate (often an insect or another small arthropod), within which they form larvae. Infected invertebrates may then be eaten by the final mammalian host or by another intermediate host that's in the food chain. The main point of the life cycle is to integrate tapeworms into the trophic web of their preferred final hosts, where worms can mature and reproduce. Transmission is always trophic, so hosts that are well fed are often infected by lots of tapeworms. Unlike urban legends, where hosts infected with tapeworms lose weight, in wild mammals usually the fattest animals carry the most worms.

Harrison's Rule

Tapeworms in mammals range from being very small, such as species of *Echinococcus* in wolves (see the introduction) in which the entire strobilus consists of three or four proglottids and the total length is about 5 mm (0.2"), to members of an order of tapeworms called diphyllobothridians. In 1915, Launcelot Harrison, an Australian zoologist who was interested in biogeography, suggested that the size of parasites is roughly proportional to the size of their hosts. Implications of this theory were that physical factors, including size, were important for the ecological distribution of parasites, and this idea eventually became formalized as Harrison's rule. Since that time, several studies of parasites including lice, fleas, roundworms, and parasitic barnacles have supported the rule,[1] but as you might expect in biology, a study of isopods (arthropods that include free-living types like sow bugs or pill bugs) found the opposite relationship – large hosts had smaller parasites.[2] So, how large are the largest tapeworms, and do they agree with Harrison's rule?

Giant Tapeworms

Reports on the sizes of organisms that we see, collect, or hunt are often exaggerated (think of your uncle's fish stories). There are unverified reports of tapeworms recovered from humans that are imposing. For example, there are reports of human tapeworms that ranged from 25 m (82') to 33 m (108'), but none of these reports provided a credible reference or specified the species of worm. In 2016, however, Li and Guo published a case study of a man who was anorexic and had abdominal pain, vomiting, weakness, and weight loss. Stool samples revealed eggs containing tapeworm larvae, and after anti-tapeworm drug treatment, the man passed a strobilus (without a scolex) of *Taenia saginata*, the beef tapeworm. This worm was 6.2 m (20') long – more than four times longer than the man. Beef tapeworms have that name because people get infected by eating raw or very rare beef – the unfortunate patient had a history of eating raw beef from which he had ingested many infective tapeworm larvae.

Other tapeworms that infect carnivorous or omnivorous mammals, especially bears and humans, are the broadfish tapeworms (*Diphyllobothrium*). These tapeworms are acquired by eating raw or poorly cooked fish that have creamy white or yellowish tapeworm larvae in their filets.

Adult worms vary from 2–15 m (6.5–49') long, with a maximum reported at 25 m (82') – longer than six humans lying end to end. These worms can grow a prodigious 22 cm (8.6") per day (almost 1 cm or 0.4" per hour), can be composed of more than 4000 segments, and live for 25 years! Broad-fish tapeworms in humans cause similar symptoms as those reported for beef tapeworm infections, except that *Diphyllobothrium* tapeworms also readily absorb vitamin B (faster than their hosts can), resulting in vitamin deficiency problems such as low red blood cell counts, tiredness, and weakness in hosts with large infections or in infections that include few but very large worms.

The Record

The record for the longest tapeworm, however, belongs to a species that was collected by a Russian parasitologist, A.S. Skrjabin, in 1967. While studying parasites of animals from Antarctica, Skrjabin found a tapeworm that was 30 m (98') long and 4.5 cm (1.8") wide. The tapeworm, a relative of broad-fish tapeworms, had a small, 12-lobed scolex with two grooves (bothria) at the end of a spoon-shaped neck, and each proglottid had from 5 to 14 male and female reproductive systems. As a result, Skrjabin named the worm *Polygonoporus giganticus* (*poly* – many; *gonoporus* – genital openings), reflecting its many reproductive systems and its giant size, although it was later renamed *Hexagonoporus physeteris*.

According to Harrison's rule, larger parasites should occur in larger hosts. This record-holding giant-among-tapeworms was found in the intestine of a sperm whale,

The Tale of the Tape: The World's Longest Parasite   145

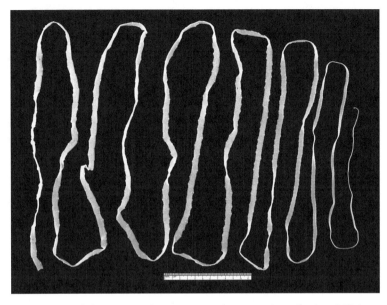

Figure 11. Adult *Taenia saginata* tapeworm, approximately 4 m (13') in length. Source: Courtesy of the Public Health Image Library, Centers for Disease Control and Prevention.

Physeter catodon, also called a cachalot, which is the largest toothed whale and largest predator on Earth. Made famous by the white whale Moby Dick, adults weigh up to 57 tonnes (125,660 lbs) and are 16–20.5 m (52–67') long. Skrjabin's giant tapeworms are almost twice the length of their giant hosts!

Sperm whales occur in all the world's oceans, including Arctic and Antarctic waters. They can dive to incredible depths (up to 2 km or 1.3 miles) in search of food, which includes mostly cephalopods (squid, giant squid, and octopuses) and fish. The life cycle of the giant tapeworm, or anything else of its biology, is unknown, but because tapeworms are transmitted trophically, it likely uses fish as intermediate hosts. Because each proglottid is about 1 mm (0.04") long, we can estimate that the entire strobila of this giant tapeworm is composed of about 30,000 proglottids. Sperm whales have no predators (except humans) and can live to be 70 years old, and each gravid proglottid of the parasite contains 80 to 100 eggs, so the lifetime production of eggs by just one giant tapeworm could easily be in the hundreds of millions (100 eggs × 30,000 proglottids × 25,550 days).

If the largest species of tapeworm occurs in the largest species of predatory mammal, is Harrison's rule supported? Tapeworms are generally small animals, and the smallest (*Echinococcus* spp.) occur in canids, such as the wolves in our introduction, but also in smaller mammals like foxes. Broad-fish tapeworms are common in large brown bears but also occur in many smaller fish-eating vertebrates, such as otters, and also in fish-eating birds, such as gulls. So, although the evidence tends to suggest that Harrison's rule probably doesn't hold true, one curious piece of evidence makes me think that there still might be something to the rule after all.

As it happens, sperm whales are also infected by another parasite, *Placentonema gigantissima*, which is not a tapeworm but a roundworm (phylum Nematoda) that lives in the uterus, placenta, mammary glands, and tissues just beneath the skin of its host. Most roundworms are small animals, no more than a few centimeters long, although ascarid nematodes are often used as examples in biology classes because, at about 20 cm (7.9"), they are relative giants in the roundworm world. However, *Placentonema gigantissima*, the titanic nematode of sperm whales, can be up to 9 m (29.5') long!

Stock's Rule

Some scientists like to scoff at biology and insist that it is not a "real" science because it has no laws (principles that are true at all times, everywhere in the universe), such as the laws of thermodynamics. They believe that biological systems can be explained by reducing them to chemistry and physics. Perhaps the closest biological idea we have that is similar to a scientific law is the theory of evolution by natural selection. So far, no evidence has been found for life on Earth that has not changed over time (evolved) due to natural selection.

The underlying principle of science is to make sense of natural observations and try to see patterns in nature that can lead to predictable outcomes. In this regard, biology is no different than any other science. Life is "messy," but biologists try to identify patterns that have explanatory power, often called rules. For instance, Bergmann's rule says that colder climate areas have larger animals, and Harrison's rule says that larger hosts have larger parasites. In deference to Carl Bergmann and to Launcelot Harrison, I wonder if it means that hosts that live in colder climates have larger parasites. While the tapeworms and roundworms found in sperm whales appear to support this idea, only time will tell if "Stock's rule" will become adopted by the scientific community.

12
Death by Raccoon

After domesticating animals, in addition to the benefits of a reliable food supply and companionship, we also acquired many infections due to our intimate contact with livestock and pets, facilitating the spread of zoonotic diseases. Today, more than 200 diseases are considered to be zoonotic, and the COVID-19 pandemic is – perhaps – the most infamous example. As our population continues to grow and more people inhabit cities, we are experiencing increasing contact with wild mammals that have been displaced from their natural habitats and live among us in urban areas, such as white-tailed deer, coyotes, and raccoons. Not surprisingly, these animals are the source of increasingly common, new zoonotic diseases – some of which can be fatal.

Zoonotic Roundworms

It used to be that most undergraduate biology students would take a lab that involved observing and doing a dissection of a pig parasite, *Ascaris suum*, as a typical example of a roundworm (nematode). For students, it had the advantages of being large (although not nearly as large as *Placentonema gigantissima*!), so many structures could be noted easily, and for cash-strapped university labs, these specimens were readily available and cost less than a dollar each.

Ascarids are some of the largest and most readily detected parasites that occur, some being 45 cm (17.7") or more in length. They live mainly in mammals, especially in carnivores and omnivores, where adults are found within the small and large intestines. The typical lab example above, *A. suum*, occurs in pigs and is closely related to a human parasite, *A. lumbricoides*. There are very minor physical differences between the two species; analysis of genes indicates that they diverged only recently and may be in the process of forming distinct, reproductively isolated species.

Humans could have first acquired *Ascaris* when we domesticated pigs and shared our homes with these omnivores, which have very similar physiologies to our own. Or, perhaps, we donated our roundworms to pigs. However it happened, archaeological studies show that, as soon as humans started keeping domesticated animals, we became infected with *Ascaris*. The infection was well known and reported by the ancient Greeks and Romans. Ascarid eggs, which are passed in the feces of their hosts, are very characteristic in size and shape, and have been recovered from human coprolites (fossilized poop) that have been dated to 2277 BC. Hippocrates treated patients for ascarid infection in the fifth century BC, but for such a large, common parasite, little was known about its biology until the nineteenth century.

In 1878, Giovanni Grassi, certainly one of the most dedicated doctors and parasitologists of all time, conducted one of the braver and grosser experiments ever attempted when he infected himself by consuming *Ascaris* eggs. These were not just any old eggs, however – Grassi had previously removed them from a corpse during an autopsy. As any good scientist would, before commencing his experiment, he examined his own feces regularly to ensure he was not already infected. Once assured he was uninfected, Grassi began his test, and about two months after ingesting eggs from the corpse's worms, he was gratified to begin to see ascarid eggs in his own fecal samples.[1]

However, what might have appeared to be a simple, direct life cycle is in fact not so simple. In 1922, another truly committed scientist, Japanese pediatrician Shimesu Koino, infected himself and a volunteer with *Ascaris*, developed a cough, and subsequently discovered small juvenile worms in his sputum. It turns out that, after swallowing eggs that enclose infective second-stage juveniles in contaminated food or water, the worms hatch in the first part of the small intestine where the adults will eventually live. But before becoming adults, the worms undergo a migration. After crawling through the intestinal lining, juvenile worms get into venules or lymph vessels, pass through the right side of the heart, enter the pulmonary blood vessels, and travel to the lungs. Here, they break out into air spaces, molt, and crawl up the respiratory tree to the pharynx, where they induce coughing. These were the stages Koino found in his sputum.

After the still-not-adult worms are coughed up and swallowed, they pass through the stomach and reach the small intestine, exactly where their migration began. Here, they will develop into adult males and females. No one knows why ascarids undergo this strange trek through the body, ending where they started, but there are some ideas. One suggestion is that they are simulating being in an intermediate host, which was once part of their life cycle but no longer exists – their one-host life cycle seems to be a recently evolved condition. However, another possibility is that the migration through host tissues allows for faster growth and larger adult body size, which the worms need to survive and to persist in the intestinal tract without being expelled from their host. Regardless, once back in the intestine, worms mature, mate, and produce eggs. After passing from the host, embryonated eggs take about 10 days to form infective juveniles.

Ascarid eggs are remarkably tough and resistant to drying and freezing, and even survive exposure to strong chemicals like the preservative formalin. Because of this resistance, many university biology classes have stopped using them as specimens in labs due to concerns that students will get accidentally infected. Currently, health authorities believe that more than one billion people in the world (especially in less developed nations) are infected with *A. lumbricoides*.

Many people will be acquainted with ascarids from their pet dogs and cats. Kittens and puppies will vomit or defecate ascarid worms (*Toxascaris leonina* or *Toxacara* spp.), which they were infected with before they were born. If their mother was infected by adult worms, juveniles can pass through the placenta to infect the young before birth; after birth, worms can be transmitted in mother's milk. Usually veterinarians recommend that puppies and kittens be dewormed, sometimes more than once. Adult pets get infected by eating parasite eggs in the environment – off-leash parks in cities are centers of infection for dogs, and playground sandboxes are hot spots for cats (which view them as litter boxes).

Raccoon Ascarids

Of course, wild mammals are also frequently infected by ascarids. For example, bears, skunks, badgers, and most carnivores are infected by different species of ascarid nematodes, but one species, *Baylisascaris procyonis* in raccoons (*Procyon lotor*), is of special concern.

Raccoons are medium-sized omnivores that occur naturally from Panama to Central America and Mexico, throughout the entire continental United States, and in southern Canada up to northern Saskatchewan (Fig. 12); they were intentionally introduced to Japan and Europe. As shown by their large geographic range, raccoons are highly adaptable mammals that can live in diverse habitats including forests, prairies, coastal areas, and even deserts, but they are now notorious for their success in urban areas. This success was

Figure 12. Raccoons (*Procyon lotor*) thrive in urban ecosystems and carry roundworms (*Baylisascaris procyonis*) that cause disease and death in other mammals – including humans. Source: David Selbert.

made abundantly clear to me when a large female raccoon got into the rafters in my father's garage in a city in southern Ontario and tipped over a can of white house paint onto his brand new Buick. Some city dwellers now call raccoons "trash pandas" because they scavenge unsecured garbage cans and cause damage to houses.

Raccoons are crepuscular (active at twilight) and nocturnal, hiding in communal dens such as tree hollows, rock crevices, brush piles, and abandoned buildings during the day. They are great climbers and swimmers. Young raccoons, especially males, leave their resident den at about a year of age and can disperse from 20 km to as much as 275 km (12–170 miles) to establish their own home range.

Because of their dispersal ability, their cosmopolitan behavior, their ability to thrive near humans, the reduction in numbers of natural predators such as wolves, and their broad diet, raccoons are expanding their range and increasing their numbers. Now, urban raccoons are often considered pests and spark calls to city animal control agents to trap and remove them.

Baylisascaris Biology in Raccoons

Other than just being nuisance animals that tip over garbage, however, raccoons are also proving to be dangerous. Raccoons are infected with adult *Baylisascaris* roundworms. A study conducted in Central Park, New York, found that 75 percent of raccoons were infected by adult worms, and there were worm eggs in 31 percent of feces that were tested.[2] Eggs are deposited by racoons in communal latrines – shared defecation sites that mark territories. A single infected raccoon may be passing 25,000 worm eggs per gram of feces. Each adult female worm inside a raccoon can produce 150,000 eggs per day, so latrines easily become contaminated with millions of eggs. Under good environmental conditions, the eggs take about 14 days to become infective, but eggs are very resilient and can remain infective on the ground for years.

Unlike other ascarids, *Baylisascaris procyonis* has a very flexible life cycle. Young raccoons get infected by ingesting eggs, while older animals get infected by eating juvenile worms that occur in a variety of intermediate hosts, especially rodents. Unlike human ascarids, raccoon worms have more life cycle options open to them and can live in the tissues of small mammals. Older raccoons tend to be less infected than young ones. Perhaps some degree of immunity develops, or possibly, young racoons get exposed to so many eggs in dens and near latrines that greater infections occur. When raccoons get infected by swallowing eggs, the worms develop in the intestinal lining, molt, and grow; they then enter the intestinal canal to become adults, where they probably consume semi-digested host food as it passes down the intestinal tract. When raccoons get infected by eating an intermediate host, the worms molt and mature to adults directly when they reach the intestine. In both cases, there is no migration through the bodily tissues of the raccoon.

Female adult raccoon worms are about 20 cm (7.9") long and tan in color, while males are similar but only about half as long. About 90 percent of young raccoons are infected, while about 45 percent of adults have the mature worms in their intestines. Young raccoons host about 45 worms each on average (although one raccoon examined had more than 1300 worms); adults are usually infected by about 18 worms. There is little tissue damage caused by *Baylisascaris* in raccoons, although extremely heavily infected animals may become emaciated because of intestinal blockage.

Many animals (more than 100 species of mammals and birds) act as intermediate hosts for *Baylisascaris*, but small rodents, rabbits, and birds seem to be particularly important. These animals share a similar habitat with raccoons, including near their dens and latrines, so they are more likely to get exposed to *B. procyonis* eggs. Because juvenile

Baylisascaris undergo migration in tissues of these hosts, infected animals suffer great damage and often die. Researchers have found that animals most susceptible to damage and death by the worms change their use of habitat just to avoid raccoon latrines. Other animals that are less susceptible do not seem phased by raccoon latrines, and some will even forage for undigested seeds there.

A Deadly Zoonosis

Unfortunately, humans can get infected with *Baylisascaris* as well, and this circumstance is when raccoon worms can become a deadly zoonosis. For example, one case report discussed a lethargic young child, with a mild respiratory cough and constant low-grade fever, who was hospitalized.[3] Suspecting a typical bacterial infection, antibiotic treatment was given, but it had little effect. Sadly, signs worsened, and it was determined that the child had mild hepatomegaly (enlarged liver), up-and-down oscillation of their eyes, and stupor. Examination of blood and cerebrospinal fluid showed a high white blood cell count (particularly eosinophils – an indicator of infection), but bacterial cultures were negative. On the third day in hospital, eosinophils continued to rise, and on day six, the little patient had cardiorespiratory failure and was placed on a ventilator. Tragically, the child died.

An autopsy revealed many white nodules (granulomas) on the surfaces of the lungs that contained nematode larvae; however, the most serious damage was seen in the child's brain. The brain was swollen, and there was severe inflammation and tissue death, as well as many "track-like" spaces. Nematode larvae were found in the cerebrum, cerebellum, and the child's spinal cord. The larvae were later determined to be *Baylisascaris procyonis*. In this case, the little child had spent time at a farm and had been seen chewing on the bark of firewood inside the farmhouse. The firewood had been collected and split outside, near a raccoon latrine. Soil in that area was heavily contaminated with *B. procyonis* embryonated eggs, containing larvae that readily became active when warmed in a lab.

After this case study was published, 14 more confirmed cases (almost all in children) were reported. Today, the prevalence of *B. procyonis* in humans is unknown, but it is undoubtedly high because we now understand that many infections are asymptomatic. A survey of blood from 150 people in Santa Barbara County, California, found that 7 percent had been exposed to *Baylisascaris*, but these people showed no clinical signs.[4] When humans or other intermediate hosts such as rodents consume eggs, the juvenile nematodes wander through tissues, causing potentially fatal damage – occasionally in humans the worms even migrate into eye tissues and cause inflammatory damage and blindness. The worms are so dangerous that a single nematode in the brain of a vole results in death, while larger hosts (such as squirrels) die with as few as five worms – no wonder these mammals have a fear of feces.

Now we know that *Baylisascaris procyonis* is a common parasite in raccoons and is significant due to its potential to cause zoonotic disease. As raccoons continue to enlarge their geographic range and are becoming more common in urban settings (they have just started to be seen in my city in northern Canada), risks of infections in people will increase. Also, raccoon worms may "spill over" into pets and urban coyotes, thus magnifying the potential danger.

Living with Danger

Raccoons can (inadvertently) kill. Their resilient parasites, adapted to survive tough times outside their hosts and to debilitate and kill intermediate hosts so that they can then get eaten by raccoons, present a problem that is likely to only increase and will be a real challenge to control, especially in cities. Improving testing and diagnosis, developing better human treatments or preventive vaccines, and perhaps treating raccoons with anti-parasite medicated baits will be necessary to control *Baylisascaris*. Great care should be taken before raccoons are intentionally translocated to new habitats to ensure only worm-free animals are introduced. In areas where raccoons are already endemic, educating people about the zoonotic dangers and encouraging urbanites to properly store garbage and to not feed the animals is important – many people think of raccoons as cute and rascally masked bandits, unaware of the dark side of raccoon ecology. As cities around the world sprawl into natural areas, and as we intentionally introduce exotic wildlife near our homes, we should expect and prepare for new and potentially deadly zoonoses to become the new normal.

13

Moths, Sloths, Tears, and Blood

The idea that nature abhors a vacuum – *horror vacui* – is a proposition attributed to Aristotle and debated by ancient Greek philosophers and early physicists, who argued about whether a vacuum or void can exist. Although modern physicists believe they have answered the question with a resounding "yes it can," the biological twist on this idea asks if there is any mechanism on Earth that some kind of living thing has not evolved in order to make or extract organic carbon compounds and energy for growth and reproduction. In this area, Aristotle may be partly vindicated, as apparently there are very few ecological niches that nature has not already filled.

Some mechanisms are very noticeable and easy to observe. For example, photosynthesizers harness solar energy and use it to cleverly take inorganic carbon in the form of carbon dioxide, which they then use to make carbohydrates, fats, oils, and proteins – plants and algae are common and abundant in every ecosystem. Herbivores consume plant molecules and break them apart to extract the carbon they need, as well as chemical energy, and predators get these resources by eating herbivores.

Parasites are everywhere and include representatives from every kingdom of life, and animal parasites have not only survived but have thrived by evolving adaptations that exploit every kind of tissue there is, as well as their hosts' food. One quirky example includes parasites that use the blood and tears of their hosts and others that cross the boundary between harmful parasites and helpful mutualists – moths. These animals show us that there are ecological niches that we never even dreamed of and yet have already been filled due to nature's biological *horror vacui*.

Justifiable Mottephobia

Mottephobia is an anxiety disorder in which people have a fear of moths. It is not as common as Arachnophobia (fear of spiders), but it still affects many people. When exposed to moths, the disorder triggers panic attacks with symptoms such as rapid heartbeat, shortness of breath, trembling, dry mouth, crying, and running away. To most of us who do not suffer from mottephobia, moths are harmless and mostly admired for their beauty. When I was a boy, giant silkworm moths such as the *Polyphemus*, the *Cecropia*, and the spectacular, lime-green *Luna* used to come to our porch lights, and I loved to see these insect titans. But to a mottephobic person, moths are morbidly frightening. The psychological condition can render sufferers afraid of being outdoors at night in summer, especially near lights where moths tend to aggregate. However, perhaps some of the fears of mottephobes are not entirely unjustified.

Moths are insects in the order Lepidoptera. All have scales on their wings that come off like dust when they are handled, and their mouthparts are adapted for sucking fluids, such as plant nectar and fruit juices, since the maxillae form an extendable proboscis – a drinking straw coiled like a spring under their head. Moths exhibit complete metamorphosis. Eggs hatch to produce caterpillar larvae that have chewing mandibles and mostly feed on plant material, so adults and larvae do not compete for the same resources. Moth caterpillars have silk glands and make cocoons within which pupation occurs and from which adult moths emerge, inflate their wings, and fly away.

To an entomologist, moths are not considered a discrete taxonomic category (one that reflects animals that are genetically related and come from a common ancestor), but generally moths are distinguished from butterflies because moths are nocturnal and have bushy antennae. There are no characteristics of moths, however, that do not occur in butterflies, and not all moths are nocturnal or have bushy antennae. There are about 250,000 species of moths, ranging in size from those with 3 mm (0.1") wingspans to giants with wingspans of 275 mm (10.8"; for instance, the tropical Hercules or Atlas moth, *Coscinocera hercules*). But the moths that deserve serious consideration from a mottephobic person belong to the genus *Calyptra* – vampire moths!

Vampire Moths

In Southeast Asia, Africa, and Russia, there are 10 species of moths in the genus *Calyptra* that pierce skin and drink mammal blood. Some feed on humans, but domesticated hoofed mammals and wildlife with no paws that can be used to groom, including thick-skinned rhinoceros and elephants, are the usual hosts. The proboscises of these vampire moths are robust and sharp, and armed with erectile barbs and hooks in sockets (Fig. 13). The tip of the proboscis is a drill, and once skin is broken, the two sides of the proboscis slide back and forth to saw through the skin, break blood vessels, and create a pool of blood.

Based on reports of human attacks, the stabs are painful and cause stinging, irritation, inflammation, and swelling. Surprisingly, however, vampire moths are not using host blood as food as do vampire bats (chapter 9), and they have many differences from insects that do, such as tsetse flies (chapter 7). First, only male moths are vampires – in most blood-feeding insects, it's the females who need blood proteins to lay eggs. Also, unlike most blood-feeding insects, vampire moths are not attracted to carbon dioxide in their victim's breath, nor do they make any anticoagulants to stop blood from clotting. Finally, they have no protease enzymes to digest blood cells or proteins; when they drink blood, they are only after one thing – salt.

Males keep very little of the salt they get from blood themselves; instead, they pass 95 percent of it to female moths in their ejaculate during mating. Females, in turn, give the salt to their offspring and replenish their own supply before

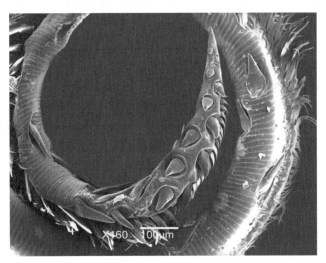

Figure 13. A vampire moth and a magnified view of its proboscis. Sources: top: Vampire moth, *Calyptra thalictri*, feeding on human thumb, photo taken by Dr. Sharon Hill; bottom: *Calyptra thalictri* Borkhausen proboscis, micrograph taken by Jennifer Zaspel.

depositing more eggs. Males use blood salt as a nuptial gift – a dowry to win over their mates.

How Moths Became Vampires

Male and female vampire moths get most of their nutrients by piercing hard-skinned fruits. This observation led researchers to analyze the evolutionary history of the family of moths that include vampires (Noctuidae: Calpinae). It seems that vampire moths evolved from nectar feeders that later added fruit piercing to their feeding regimen. Mouthpart adaptations needed for piercing tough-skinned fruit were the same needed to penetrate mammalian skin, so vampire moths were predisposed to be able to take blood.

Nevertheless, not all members of the vampire moth family use blood – some only pierce fruit, although the structure of their proboscis would let them take blood. It turns out that the major difference between blood-feeding (or, more properly, blood-using) moths and those that don't use blood is the number of sensory organs on their antennae – blood-users have fewer. At first, this difference seems strange, because you would think that the blood-feeding moths would need more acute senses to find their hosts and locate blood vessels at the skin's surface. However, researchers suggest that, with poorer senses, blood-feeding moths are not repulsed by the stink of mammals – in a sense they "hold their noses" to drink blood and get the salt they desire.

Crying over Vampires

While normal human blood typically has about 5 mg/ml (0.005 oz/fl. oz) of salt, tears have quite a bit more at 6 mg/ml (0.006 oz/fl. oz), so it is not surprising that many moths from a variety of families have taken up the habit of drinking mammalian tears, a mode of feeding called lachryphagy. Notably, one species in the vampire moth family only drinks tears – the tears of birds in Madagascar, since on Madagascar there are no native hoofed mammals to use as hosts. This moth, *Hemiceratoides hieroglyphica*, uses its piercing proboscis to penetrate the nictitating membranes (the thin, transparent membranes that can be drawn across the cornea to clean and moisten the eyes) of bird eyes and drinks their tears.

Most lachryphagous moths feed on large mammals, especially deer, bovids, horses such as zebras, and even elephants – but they also drink from humans. So far, more than 100 tear-drinking moth species have been recorded, with 23 that can use humans. Moths land on the host's face, move around, and probe other openings (like nostrils) before drinking tears – they do not seem to have the ability to specifically detect host eyes. The irritation induced by the moth's proboscis causes excess tearing and, therefore, a larger harvest of salt.

Tear feeders probably evolved the habit independently several times and likely started as "puddlers" – moths that get salt from mud puddles and organic debris like dung. Salt is a valuable resource in short supply, especially in tropical areas with heavy rainfall where it gets leached from the soil.

Like vampire moths, tear feeders are usually males, so salt collected is likely donated to females as reproductive gifts. One species, however, *Lobocraspis griseifusa*, is an exclusive tear feeder, and both males and females get nourishment, in addition to salt, from lachrymal secretions.[1] This very specialized moth has a short proboscis and during tear-feeding holds its wings upright. Consequently, many moths (up to 12) can crowd around one eye (usually a cow) to drink. Unlike other tear-drinking and vampire moths, this species has proteases and can digest proteins from its lachrymal meals. Considering these discoveries, it seems that mottephobics have valid evidence to support their fear. But before we condemn moths altogether, there are some that are probably responsible for keeping their hosts alive – sloth moths.

Sloth Moths

Sloths are incredible and fascinating mammals. They belong to the order Xenarthra (*xen* – different; *arthra* – joint), which includes mainly ground-dwelling animals like armadillos and anteaters. All have a unique bony process on their lower back (lumbar) vertebrae, which gives extra bracing to their sacrum and provides a strong, muscular base for the ground-dwelling members of the group to dig and burrow but also allows sloths to hang upside down from branches in the canopy of the rainforests where they live.

Sloths occur in two families, Megalonychidae (two-toed sloths) and Bradypodidae (three-toed sloths), and both are tree-dwelling leaf-eaters that spend their sleepy lives in the canopy of tropical rainforests in Central and South America. This lifestyle is rare for mammals – only 10 species in the world are known to use this specialized ecological niche. It may seem surprising because the biomass of rainforest leaves available to eat is immense, but although food is abundant, it is very challenging to use. Rainforest trees are attacked by countless numbers of insects and other herbivores, so they defend their leaves physically with spines, tough fibers, and gummy resins like latex, and chemically with a huge variety of toxins, including alkaloids, phenolics, tannins, cardiac glycosides (which cause irregular heartbeats in mammals), and others. Even when animals can digest this toxic, high-fiber diet, they do not get many essential nutrients such as proteins and lipids.

Sloths have many specializations for living in the canopy and eating toxic leaves. They are small (4.5–6.5 kg or 10–14 lbs) but long-lived (30 years), and they spend most of their life hanging upside down from branches, using their elongated limbs and long claws as hooks. Their rigid vertebral column and small skull reduces weight and assists with their life in trees. Like ruminants such as cows, they have guts that include sacs and very long intestines filled with protozoans and bacteria that can digest plant fiber, which makes up almost all of their diet.

This diet is difficult to break down. Sloths have the slowest rate of digestion of any mammals and get so little caloric value from their food, about 110 cal/day (the equivalent of one small baked potato), that they have the slowest metabolic rate known for mammals. To save the precious energy that they can extract from their diet, they do not thermoregulate, so they must live in the warm tropics. Of course, their world-famous slow locomotion (a common Spanish name for sloths is *los perezosos* – the lazy ones) saves energy and is the basis for their English name.

Three-toed sloths (*Bradypus variegatus*) are more specialized for handling a diet of leaves than are two-toed sloths (*Choloepus hoffmanni*). Two-toed sloths have more diverse diets, including some fruits and animal matter, and have larger home ranges, while three-toed sloths can spend their entire lives eating leaves in the canopy of just two or three neighboring trees. The arboreal, leaf-eating life of sloths is a very constrained ecological niche, and not many mammals have evolved to use it. You might think that their vegetarian life in the rainforest canopy would reduce the chances of sloths getting infected by parasites, but even they get infected with intestinal worms, including tapeworms, acquired from eating mites, and roundworms, which they get by eating insects. Besides the endless abundance of leaves for food, however, there is another major advantage of this lifestyle.

There are far fewer predators of adult sloth-sized animals in the canopy than occur on the rainforest floor. Large, but rare, harpy eagles can kill sloths in the treetops, but the forest floor has jaguars, ocelots, margays, short-eared dogs, anacondas, caimans, and of course human hunters. Despite the little time that sloths spend on the forest floor, however, most sloth mortality occurs there, and yet sloths (especially three-toed sloths) display a very unexpected behavior: they slowly climb down tree trunks to the dangerous forest floor in order to defecate. Besides being risky, this activity is energetically expensive, using about 8 percent of a sloth's daily energy budget.

Why would sloths undergo this reckless, expensive behavior just to defecate? The answer is moths.

In the fur of sloths live moths belonging to the genus *Cryptoses*. These insects belong to the family Pyralidae, which are called snout moths or grass moths. Two-toed sloths have on average about five moths living in their fur, while three-toed sloths have about three times more. These moths are host specific, occurring nowhere else in the world except in the fur of sloths. When sloths make their perilous journey to the forest floor, they scrape a hole at the base of their home tree, defecate, and use their stubby tails to cover it with leaf litter. As they defecate, female moths leave the fur to deposit eggs in the dung. After hatching, moth caterpillars eat sloth dung, pupate there, and newly emerged male and female adult moths fly upwards into the tree canopy to find and colonize the resident sloths.

The fur of sloths is an ecosystem unto itself. Sloth hairs have crosswise cracks that collect water from rain and mist. Green algae (*Trichophilus* spp.) grow hydroponically in the fur, using the abundant unfiltered sunlight of the rainforest canopy for photosynthesis. As a result, sloth fur is a luxurious green color, which provides camouflage from sharp-eyed eagles; the algae garden also provides a substrate for fungi that live there too. This microbial biomass (algae and fungi) is not trivial – it accounts for about 3 percent of the weight of a sloth. Carrying the garden around with them exerts an energetic cost, so there must be a payoff. When sloths groom their fur, they ingest a salad of algae and fungi, getting as much as five times more lipids and proteins than they get from their regular leafy diet.

But what about the moths – are they parasites or mutualists? Moths are portals for nutrients, eating sloth skin secretions and some algae, while their wastes are fertilizer for the garden, as are decaying moth bodies when they die in the fur. Moths also carry some fertilizing nutrients from the sloth dung below when they colonize a host. As complex and hard to believe as this ecological system is, sloths promote moth infestations of their fur and cannot thrive without them. The lives of sloths and their moths are locked together in a beautiful symbiotic relationship.

This amazing mutualistic syndrome is critical to the survival of sloths. Natural selection has led sloths to trade the risk of predation for cultivating moths that contribute to their limited nutrient resources. Although many mammals, including some humans, are mottephobes – rightly fearing blood-feeding and tear-drinking parasites – sloths are mammals that are mottophilic and have come to embrace their moths to the extent that they now live in symbiosis.

The relationships between moths and sloths beautifully illustrates that biotic interactions can take completely different paths over evolutionary time. In some cases, both partners benefit from the symbiosis (mutualism), while in others the relationship becomes asymmetric and one partner benefits at the other's expense (parasitism). Trying to establish the chain of random events that leads to one outcome or the other is a fascinating problem.

Regardless of whether we will ever discover the precise mechanisms and events that have led sloths to migrate to the jungle canopies, developing their garden along with their moths, one thing is plain: even though the countless number and range of ecological niches that are out there is truly astounding, biologists rarely if ever find one unoccupied. Nature really does abhor a vacuum.

14
The Manchurian Parasite

Parasites have evolved a profusion of physical structures to successfully find and hold on to hosts, as well as special body shapes and surfaces to extract and absorb food more efficiently. It should come as no surprise, then, that they have also found ways to manipulate hosts by using chemicals. Hormones can be affected so that hosts reproduce at inappropriate times or try to mate with inanimate structures, and brain messenger chemicals can be altered in order to cause hosts to lose their innate fear of predators or to move into habitats that are usually avoided. We are getting more and more evidence that single-celled protozoan parasites (those that cause diseases such as malaria) alter host behavior, but some animal parasites have taken this evolutionary opportunity to influence behavior and really run with it, even going so far as to evolve the ability to control their hosts' minds.

In 1959, Richard Condon wrote *The Manchurian Candidate*, a chilling political thriller about a man who is brainwashed to be an assassin. This novel spawned a movie and subsequently many stories about the potential for mind control. For humans, these ideas are thankfully not something we need to worry about (although with the emergence of novel zoonoses, who knows what the future might hold), but in nature some parasites actively use mind control to turn other animals into zombies, making them lose their self-control and unable to do anything but the bidding of their parasite puppet masters.

Thorny-Headed Worms

The animal parasites that have refined the ability to change the normal behaviors of hosts are acanthocephalans, commonly called thorny-headed worms (*acanth* – thorn; *cephala* – head). These worms are undeniably strange, highly modified creatures that are all parasites. Adult males and females live in vertebrates, mainly in fish but also in reptiles, birds, and mammals. There are only about 1000 known species, and they always live in the digestive systems of their hosts, particularly the small intestine, where they use a retractable, finger-like proboscis to embed in the intestinal lining (Fig. 14). The proboscis is armed with rows of hard, recurved hooks (the "thorns"), which anchor the worms and prevent them from being flushed out of their hosts. At the anchor sites, the worms can cause localized damage to the intestine, and there is evidence that each worm moves around and stabs the intestine several times, causing multiple wounds.

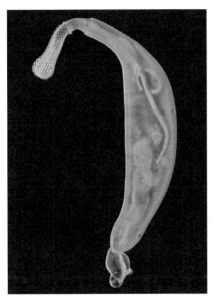 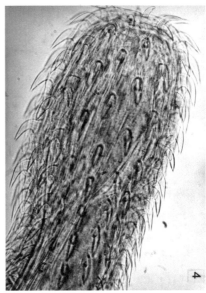

Figure 14. An adult male acanthocephalan and a proboscis armed with hooks. Sources: left: iStock.com / Sinhyu; right: Cath Wadforth / Science Photo Library.

Over evolutionary time, thorny-headed worms have lost their mouths and digestive systems and instead feed on semi-digested host food that they absorb through their body wall, which functions much like the tegument of flatworms but evolved separately and is more closely related to the non-living cuticle of arthropods. It is a good example of convergent evolution, where unrelated organisms evolve functionally similar adaptations to solve a problem, in this case, absorbing nutrients. This feeding mechanism has led to an interesting observation – thorny-headed worms accumulate and store environmental pollutants, particularly heavy metals like lead and cadmium, more intensely than do their hosts, so acanthocephalans can be used as early warning biomonitors to detect pollution problems in ecosystems.

Adults are usually about 1 cm long (0.4"), and at the rear end of males, there is a copulatory bursa – a flanged extension of their body that can latch onto a female during insemination – and also cement glands. The secreted cement is called a copulatory cap and is used to seal over the genital opening of a female for a few days after copulation. This cap plugs a female's vagina and ensures that she cannot reproduce with other males – a worm's interpretation of a chastity belt. Occasionally, male worms are found with copulatory caps on their surfaces, indicating that other male worms have tried to mate with them, or perhaps one male has tried to block the reproductive system of another. In the intestines of ringed seals (*Phoca hispida*), large male thorny-headed worms (*Corynosoma magdaleni*) were found clustered around unmated females (females without copulatory caps), suggesting that there can be intense sexual selection in acanthocephalan populations.

After mating, female thorny-headed worms have a special organ, the uterine bell, that sorts mature embryonated eggs from slightly smaller immature eggs and allows only mature eggs to be passed into their host's intestine. From a single copulation, a female can produce thousands to millions of embryonated eggs, numbers that are necessary in order for a few worms to successfully survive and make it to another host.

How to Infect Hosts

All acanthocephalans have life cycles that involve at least two hosts. The first host is invariably an arthropod, frequently an insect or crustacean. These hosts get infected by eating embryonated eggs that have been deposited by a host infected with adult worms; the more eggs released into the environment, the better the chances are that some will get eaten. The embryos, called acanthors, look like elongated pears with six or eight bladelike hooks at the front end. Acanthors can survive inside shelled eggs for months, waiting to be eaten by an arthropod. Eggs of a species that infects wild and domestic pigs (*Macracanthorhynchus hirudinaceus*) can survive freezing and drying, and remain alive and infective for more than three years in soil.

After getting eaten by an arthropod, the acanthors use their proboscis hooks to dig through the intestine and take up residence in the host's body cavity. Here, thorny-headed worm larvae absorb nutrients, grow larger, and become the next larval stage – an acanthella. Finally, the acanthella develops rudimentary organs, which again include an inverted spiny proboscis. The larva is often covered by an envelope or capsule, which may protect it from the arthropod's immune responses. At last, when the larva is developed and ready to infect a vertebrate host, it is called a cystacanth.

Suicide by Predator

This stage in the life cycle is where thorny-headed worms use host mind control. In the late 1960s, parasite ecologist John Holmes and a graduate student were working at a shallow lake in central Alberta when they noticed that many of the resident amphipod crustaceans (*Gammarus lacustris*, commonly called scuds) were clinging to floating plant material.[1] When the scientists came out of the lake, some scuds were clinging to their hip waders, dip nets, and arm hairs and remained attached even if disturbed. Normal scuds are gray-green camouflaged scavengers that live at the murky bottom of lakes and rapidly disperse if disturbed, so these weird scuds were behaving very abnormally. Besides being at the water surface and clinging, something else was different about them – they had bright orange spots on their top surface, clearly visible through their shells.

Later in the lab, it was discovered that the orange spots were cystacanth larvae of a thorny-headed worm, *Polymorphus paradoxus*, a parasite that occurs in the small intestines of beavers, muskrats, and mallard ducks. Could it be possible that the parasites were causing behavioral changes to the scuds, and if so, why? Experiments in the lab showed that scuds infected with acanthocephalans spent more than 80 percent of their time in light (versus 1 percent for uninfected) and 70 percent at the water surface (versus 1 percent for uninfected). Moreover, if the light source was changed, infected scuds would move toward light instead of avoiding it, which is their natural response.[2]

In another experiment, infected and uninfected scuds were put into an aquarium with pieces of floating weeds and wood, which confirmed field observations. The only scuds ever found clinging to floating debris were infected, and about 60 percent of the infected amphipods were found clinging at any one time. The crustaceans used their claws to cling tenaciously; even when the weed or wood was shaken, the scuds held on. Parasite-infected hosts would skim along the surface of the water and cling to any floating material they found.

The researchers then did tests where they presented 75 infected and 75 uninfected scuds to one or two well-fed mallard ducks. Telling the infected and uninfected scuds from each other was easy because of the bright orange spots that the infected scuds sported. After six repeated tests, 68 percent of infected scuds were eaten, while 16 percent of uninfected amphipods got eaten, meaning parasite-infected hosts were over four times more likely to get eaten by mallards. Subsequent tests showed similar results for muskrats (*Ondatra zibethicus*). The thorny-headed worms were exerting mind control over scuds, causing them to change their normal, adaptive behaviors and to commit suicide by predator!

Duped scuds were facilitating the parasite's transmission into vertebrate hosts – especially, in this case, those that fed at the water's surface. But shouldn't muskrats and other hosts learn to avoid these orange-spotted infected zombies? It appears that the costs of getting infected by thorny-headed worms in these hosts must be less than the large amounts of extra calories they get by dining on easy-to-catch prey. The acanthocephalans must not cause a lot of damage to the intestines of these hosts before they complete their life cycle. Therefore, there would be little selection pressure to avoid infection. For muskrats, beavers, mallard ducks, and *Polymorphus*, this arrangement is a win-win situation – the scuds, however, are big losers.

Later, Holmes discovered another related species of *Polymorphus*, *P. marilis*, which also induced mind control of scuds but with a twist. In this case, the worms were causing infected scuds to go to a completely different habitat, where they got eaten by diving predators like scaup ducks.

Evolution by natural selection had led to two divergent but successful strategies that incorporated mind control.

Suicidal Cockroaches

The startling discovery that *Polymorphus* worms caused altered behavior led biologists to wonder if this mind control was just a weird but unique phenomenon. Other species of acanthocephalans are larger than *Polymorphus*, have terrestrial rather than aquatic life cycles, and occur in a variety of mammals. For example, rats worldwide are infected with *Moniliformis moniliformis*, large acanthocephalans with males measuring about 8 cm (3.1") long and females 20 cm (7.9"). Their proboscises are long, cylindrical, and armed with 12 rows of 10 hooks, so one of these large worms has an arsenal of 120 hooks. Like other acanthocephalans, they use these hooks to attach to their hosts' intestines. After being fertilized, adult females release mature eggs, which are excreted from infected rats.

On the ground, eggs get eaten by intermediate hosts – often cockroaches. Electrolytes in roach intestines stimulate acanthors to break down eggshells, and the released larvae use their hooks to pierce the insect's midgut. In about 55 days (depending on temperature), cystacanths will have matured in the roach's body cavity, are enclosed in membranous cysts, have their proboscis withdrawn into a pouch, and are ready to infect rats. Rats, the final hosts, get infected by eating roaches.

Researchers have discovered that *Moniliformis* causes behavioral changes in cockroaches, making them more susceptible to predation by rats.[3] For instance, infected roaches move toward light and move more readily than uninfected insects, and the altered responses did not depend on infection size. Also, when tested on a roach-sized running wheel, infected roaches spent more time running and moved slower – particularly at night. This altered behavior would make infected roaches more vulnerable to rats, especially because these changes tend to occur nocturnally, just when rats are foraging.

Cockroaches have a very instinctive, programmed escape response, and it too was affected by acanthocephalans. When roaches were infected, they took longer to move when stimulated and moved shorter distances than uninfected hosts. While 90 percent of roaches tested had escape responses when stimulated, only 55 percent of infected roaches reacted normally. Even with small infections, roaches took longer to escape, required more stimulus to try, and escaped less often.

One complaint some biologists had about altered behavior studies was that the apparent mind control might not be the result of evolution – it might not be mind control at all – but rather simply due to damage that parasites were causing in hosts. So, just in case these altered behaviors were due to pathology or reduced roach stamina, researchers showed that roaches infected with thorny-headed worms performed other behaviors normally, such as responding to pheromones (chemicals released by roaches to communicate with each other).[4] Also, the velocity of escape movements between infected and uninfected roaches was the same. There was no sign of any damage in infected insects, and they lived as long as uninfected cockroaches – if they didn't get eaten by rats, of course. It is now clear that the altered behaviors induced by

Moniliformis are specifically those that would cause roaches to get eaten by rat hosts, within which the parasites mature and reproduce. This mind control is no accident!

The mechanism by which thorny-headed worms exert mind control on their intermediate hosts is both complex and impressive. Rather than causing a novel behavior, acanthocephalans induce hosts to express pre-existing behaviors inappropriately – at the wrong time or in the wrong place. For example, the clinging response of scuds to weeds is behavior they normally use during reproduction – amphipods grasp each other during mating. Normal host responses are affected because worms secrete neurotransmitters – brain chemicals such as octopamine or dopamine – to ramp up behavior (for example, clinging) or serotonin to dampen it (for example, inhibited escape response).

The Host Fights Back

Given that all thorny-headed worms so far tested cause altered behavior in their arthropod intermediate hosts, and the mind control leads to suicide, evolutionary theory would suggest that crustaceans, isopods, millipedes, and insects should have evolved defenses over time. For example, the integrity of the envelope that forms around a cystacanth larva in an arthropod host's body cavity is critical to its survival. In the wrong hosts, acanthocephalan larvae get encased and destroyed by arthropod immune reactions, such as melanization (the deposition of black melanin pigment). In addition, arthropods have evolved another method of defense, countering altered behavior by adjusting their own behavior to impair parasites: fighting fire with fire or rather, in this case, with cold.

The development of *Moniliformis moniliformis* in cockroaches from egg to infective cystacanth takes a long time – over 50 days at 27°C (80°F). Normal acanthocephalan development only occurs between 24°C and 32°C (75–90°F). Outside this temperature range, retarded development, abnormalities, encapsulation, and melanization of parasites occurs. When cockroaches are given the chance to choose their preferred temperature, scientists found that parasitized roaches spent more time than uninfected roaches in the cold.[5] The colder temperatures greatly affect worm development, which in turn negatively affects successful transmission of worms to rats.

Other tests indicated that, although infection had no effect on cockroach survival in temperatures ranging from 22°C to 31°C (72–88°F), *Moniliformis* did impact the number of cockroach offspring produced at higher temperatures.[6] By choosing to live in cooler conditions, roaches could avoid one of the negative impacts of being a host. Cockroaches can fight against worm infections by using behavior and choosing habitats that are unfavorable to parasites.

Amazing Acanthocephalans: Where They Came From

Although they represent a small, mostly unknown group of parasites, thorny-headed worms are amazing. As adults, they tend to have a wide range of hosts within which they can survive; for instance, *Polymorphus paradoxus* can infect beavers, muskrats, and even mallard ducks. Large terrestrial acanthocephalans, like *Macracanthorhynchus hirudinaceus* and *M. ingens*, infect wild and domesticated pigs, raccoons, and humans. They can be extremely common and reach sizes of 45 cm (17.7"). The largest reported species of thorny-headed worm, *Nephridiacanthus longissimus*, occurs in aardvarks and is almost 1 m (3.3') long. But acanthocephalans have a longer evolutionary history in their arthropod intermediate hosts than in their vertebrate definitive hosts, and they have therefore had longer to evolve wonderful specializations such as mind control in arthropods.

Thorny-headed worms are well-adapted parasites with lots of special features, but analysis of genes and many physical features show that they shared a common ancestor with small, nondescript aquatic animals called rotifers. In fact, the relationship is even closer than sharing an ancient ancestor – acanthocephalans are likely a highly modified type of rotifer. If you sample the water in any stream, pond, river, or lake, you will see that rotifers look nothing like thorny-headed worms; they are small (about 1 mm or 0.04"), ciliated, aquatic, free-living animals that lead quiet lives attached by their "toes" to algae and weeds. They have wonderful little jaws, surrounded by two wheel-like lobes covered in cilia. As the cilia beat, they drive a current of water that has bits of dissolved organic matter and bacteria into the mouth and jaws of the rotifer. However, closer inspection gives us some clues about their relationship with thorny-headed worms. Both have cement glands (rotifers use cement to attach their toes to plants), special body cavities called pseudocoelomes, body surfaces made of tissue with many cell nuclei, and other subtle features that confirm their close affinity. The intimate host-parasite relationship of acanthocephalans with arthropods gives us insights into how rotifers might have become parasites and how these little mild-mannered, free-living animals became masters of manipulation.

It appears that a small, free-living rotifer that attached to the outside of marine crustaceans, such as marine amphipods, eventually went inside and over time became an endoparasite, perhaps to avoid predators. At first, mature adult parasites would have resided in the body cavities of amphipods and resembled cystacanth larvae. They would have lived by absorbing nutrients through their body walls. Much later, vertebrate predators of the infected crustaceans would have been added as another host in the worm's life cycle. Diversification of arthropods into terrestrial habitats, likely by animals such as sow bugs, pill bugs, or roly-polies found in marine and terrestrial ecosystems, would have led to diversification of thorny-headed worms. As parasites that have lived integrated lives with arthropods, natural selection would have led acanthocephalans to eventually evolve the ability to secrete neurotransmitter proteins in order to control their arthropod hosts' behaviors, especially as it related to predators.

Great Minds Think Alike

Sometimes natural selection stumbles across the same solution to a selection pressure problem in very different organisms. For instance, both sharks and dolphins have streamlined, torpedo-shaped bodies – great for swimming fast in order to catch quick prey like fish and squid. This adaptation was not handed down from a common ancestor because sharks (cartilaginous fish) and dolphins (mammals) last shared a common ancestor more than 400 million years ago. The independent evolution of a common feature in different organisms is called convergent evolution. Features that are the same in distantly related animals speak to a similarity in lifestyle and habits (their ecology) rather than serve as clues about being evolutionary relatives.

One example of convergent evolution for parasites involves how to get offspring from one suitable host into another – transmission. Convergent evolution has resulted in many different kinds of parasites developing devious ways of tricking hosts to facilitate transmission, and altering host behavior is just one. Some single-celled parasites are masters of manipulating host behavior, but perhaps the best at manipulation are parasites that live inside cells in the digestive tracts of cats – *Toxoplasma gondii*.

Inside cats, the parasites divide and produce resistant structures called oocysts that get passed when a cat uses the litter box. *Toxoplasma* doesn't normally cause much trouble for cats, although they can be a source of diarrhea for kittens. When the cysts, which are so tiny that they can be picked up by air currents, get eaten by any kind of warm-blooded animal, they form cysts in tissues (muscles and nervous system tissues are common locations). Now the fun begins. *Toxoplasma* evolved in wild cats and has found ways of changing the behavior of typical prey animals (like rodents) so that they are more likely to get eaten by cats. Experimental studies have demonstrated that *Toxoplasma*-infected rodents become attracted to cats – in fact, they lose their innate fear and offer themselves up as food.[7]

The mechanism *Toxoplasma* uses to manipulate hosts is still under investigation, but most studies point to its ability to make, or force host neurons to make, the neurotransmitter dopamine.[8] Dopamine in humans affects how we feel pleasure but also our ability to think rationally, to focus, and even affects what we find interesting. In rodents, it may make rats attracted to cats. Humans can also get infected with *Toxoplasma* – some estimates claim that two billion people around the world are infected. This prevalence led some researchers to question whether *Toxoplasma* can affect human behavior – after all, our ancestors in Africa must have occasionally been at risk of attacks by big cats, so there would be a payoff for the parasite using mind control in humans too. Many studies have correlated exposure or infection with *Toxoplasma* to human behaviors, such as increased risk taking, more frequent automobile accidents, more extroverted behaviors, and other behaviors.[9] This idea is certainly interesting – or maybe that's just my *Toxoplasma* talking.

Convergent evolution has led other mammal parasites to use mind control as well. *Dicrocoelium dendriticum* (a parasitic flatworm fluke) occurs in the bile ducts of many mammals that eat grass – sheep, cows, goats, pigs, and deer. Fluke eggs are passed from the mammals and, if eaten by snails, develop to larvae called cercariae. These seem to irritate and clog the "lungs" of the snails, causing them to release slime balls that include hundreds of cercariae. The slime attracts ants. When ants eat the slime balls, most parasites will penetrate the gut and develop to infective stages in the body cavity of their hosts. One or two, however, will encyst around the brain (subesophageal ganglion) of the ant, and then normal ant behavior changes. Whereas most ant workers will forage during warm summer days and retreat to their nest during the cooler nights, *Dicrocoelium*-infected ants that are gnawing on grass will lock their jaws on a blade of it and stay there for the night. The parasites tell the ants to lock their jaws as it gets cooler and to stay on the grass. Most grazing hosts feed in the evening and early morning, and along with the grass they eat, the mammals will consume parasite-infected ants. A related parasite infects carpenter ants and uses mind control to make their hosts active in bright sunlight – exactly when they are exposed to foraging birds like robins. One thing that all of these puppet-master parasites have in common, whether they be single-celled parasites, thorny-headed worms, flatworms, or others, is that they are passed from host to host trophically by predation – in other words, by getting eaten. Altering normal host behavior to make them easy prey is an example of convergent evolution.

Could Parasites Control Whole Ecosystems Too?

Through the years, many ecologists have studied predation and how it can structure food webs and entire biological communities. Due to the discovery of parasites using altered host behavior, we must now realize that, hidden behind a curtain in the background, parasites are using methods like mind control that may affect larger scale biological processes. In nature, parasites may be affecting which animals get eaten, how many are eaten, and even when it happens. Parasites may be hidden puppet masters, working behind the scenes to control trophic interactions and to structure whole biological communities. Therefore, in a way, parasites are now exerting mind control on biologists – forcing them to change the way they think about predation, energy flow, biological diversity, and even entire ecosystems.

15
A Ghost of a Chance

The history of life on Earth is, ironically, largely about death. Examination of the fossil record and other evidence of the past shows that far more species of organisms have lived, died, and gone extinct than currently exist today. Biologists have long wondered if mass extinction events, such as those that marked the end of the Permian and Cretaceous periods, occur in a predictable pattern or if they are random events. On a smaller scale, conservation biologists try to measure the vulnerability of particular species or members of particular ecological niches. For instance, are large predators more prone to extinction than small fruit eaters? Do they need special protection in order to try to preserve them? Some biologists have argued that animals that are very specialized for exploiting certain habitats and have unique adaptations that make them successful under particular conditions are more likely to suffer extinction.

One such specialized species occurs in the boreal forests of North America: moose. They browse on aspen and willow, silently move through dense vegetation, and thrive in extreme cold with deep snow. Yet in spite of the wonderful adaptations used by moose to exploit the boreal forest, could they be marked for extinction? If so, the fatal factor may be a specialized parasite, aided by our alteration of the climate.

Ghosts

Nowadays, trappers are often chastised because of society's distaste for wearing fur, but people who make their living due to their extensive knowledge of wildlife often care deeply about preserving nature and conserving mammals – after all, their livelihood depends on it. Thus, it is not surprising that trappers and Indigenous people were among the first to report the spectral presence in northern forests of "ghost moose."

Moose (*Alces alces*) are among the largest hoofed mammals in North America. Bull moose can weigh more than 600 kg (1320 lbs) and carry a massive rack of palmate antlers that are close to 2 m (6.6') wide and weigh up to 18 kg (40 lbs). Members of the deer family, they are wonderfully adapted for living in northern forests. Their long legs, splayed cloven hooves, and dew claws let them navigate obstacles, move through muskeg, and travel, seemingly effortlessly, through deep snow. They are muscular athletes that can run up to 55 km (34 miles) per hour or move silently through forests if they so choose.

Moose are the best-adapted deer for dealing with the varying snow conditions and the cold temperatures of high latitudes. Their large size helps to conserve body heat, and their long legs, which would seem to be a detriment to conserving heat but a benefit to moving in snow, have a heat exchange between veins and arteries. Similarly, there is a counter-current heat exchanger in their long noses to reduce heat loss. Moose change their activity levels and metabolism to easily deal with the cold northern climate; their metabolism is not affected by cold stress unless the temperature drops below minus 40°C (–40°F). However, perhaps the most important adaptation they have is their insulative winter coat.

The winter coat of moose consists of guard hair and underfur. Guard hairs are long (more than 25 cm or 10" on the back near the shoulder hump), with large, air-filled centers, and are brown, dark brown, or black, depending on moose age, social rank, and sex. The woolly, lighter colored underfur is shorter (about 2.5 cm or 1" long) and twisted. It grows during the summer and provides most of the coat's insulation in winter. The winter coat traps insulative air between the moose and outside; it is shed in spring, giving moose a conspicuous "ratty" look, with tufts of unshed winter coat.

The Cause of Ghost Moose

Ghost moose are white and appear at the end of winter because they have groomed and damaged extensive areas of their dark coats, exposing their white undercoats (Fig. 15).

Figure 15. A ghost moose. Source: Norman Pogson / Alamy Stock Photo.

This excessive, uncontrolled grooming is caused by infestation with winter ticks, *Dermacentor albipictus* (Fig. 16). Ticks are arthropods that are all external blood-feeding parasites of vertebrates, other than fish, and their mouth parts include a pair of appendages called pedipalps that fuse together into a cone-shaped structure (a hypostome) in order to jab their host. There are also saw-like jaws, called chelicerae, that cut through skin.

As ticks cut into skin, they shove their hypostome into the wound. Backward-facing chitinous teeth on the surface of the hypostome anchor ticks to the moose's skin, and winter ticks, like other "hard ticks," also secrete a liquid cement that hardens and secures the parasites to the surface of their

Figure 16. A blood-engorged winter tick, *Dermacentor albipictus*. Source: © Brad Steinberg.

host. Once securely attached and with their head embedded into the host, the parasites suck up blood and lymph from lacerated tissue pools. The ticks secrete a cocktail of chemicals from their salivary glands that stops blood from clotting and also makes the wound less painful so that hosts will ignore them until they become attached. Some hard ticks, but not winter ticks, cause paralysis of their hosts, which prevents hosts from even trying to remove them.

Other than eggs, all life cycle stages of moose ticks (larvae, nymphs, and adults) suck blood. Larvae (called "seed" ticks) have six legs, unlike the other stages that have eight, and they attach to moose starting in late September. Seed ticks feed until October, then molt to slightly larger eight-legged nymphs, which in turn feed from December to February and then molt to still larger adults. Adults feed until March or April, mate, and then blood-engorged ticks drop off their moose host onto the forest floor to lay eggs in May and June. Thus, moose are infested with winter ticks from September to late March – seven months of the year. Blood-engorged female winter ticks, about the size of our thumbnail, have folds in their leathery cuticles that allow them to swell like water-filled balloons.

On the ground, engorged female ticks are so filled with blood that their little legs cannot move their bloated bodies. Near where they have dropped, females lay clutches of eggs in the leaf litter and die. Each female lays about 1000 eggs, and the rate of development of seed ticks in the eggs depends on environmental conditions, especially temperature. After seed ticks hatch, they use their six walking legs to crawl up into vegetation, preferring to climb as high as the chest level of moose. Clusters of hundreds of seed ticks can be seen near the ends of twigs and branches, and the little ticks form ball-like masses to retain moisture. The timing of when ticks form clusters coincides with the time when moose are most active and moving through the forest – during the rut.

Stimulated by passing potential hosts, vibrations, odors, heat, and carbon dioxide, seed ticks rear up and wave their forelegs in a "questing" behavior, although they have also been seen jumping, trying to latch onto a moose in a behavior called "tiddly-wink flips." Once on a host, larvae crawl under the fur to the warm skin below and start to feed.

In Alberta, all moose become infested with winter ticks every year they live. Most moose will be infested with more than 32,000 ticks (more than 1 tick for each cm^2 or 0.4"^{2} of skin), and infestations of more than 50,000 are not uncommon. Other hoofed mammals get infested too but not with as many ticks or as frequently as do moose – winter ticks are definitely moose specialists.

The ticks cause lots of problems for moose. Adult ticks are voracious feeders, consuming two to three times their engorged weight in blood, meaning that a typical bull must replace about 5.5 liters of blood (1.2 gallons, or about 17 percent of their total blood volume), cows 3.2 liters (0.7 gallons, or 11 percent), and calves about 8 liters (1.8 gallons, or 58 percent) during winter and early spring. Not surprisingly, moose experience anemia, low serum protein levels, and reduced fat stores, and there is inflammation and ulceration on the skin, which becomes itchy. If those problems weren't bad enough, winter ticks can transmit microbial infections, and because of the burden of so many attached ticks, moose groom excessively.

Moose Defenses Are Expensive

As a defense, moose may try to avoid getting infested by winter ticks. In an experiment where a captive moose that had previously been infested was offered food pellets with or without clumps of seed ticks, the animal chose the uncontaminated food.[1] Although almost all moose do get infested in the wild, this behavior might help to reduce the level of infestations. Once infested, moose use oral grooming (licking and biting fur), scratching using their rear hooves, and rubbing their hides against trees and shrubs to try to remove ticks. The itchy irritation of ticks moving on the hide also causes moose to shake their heads and bodies, and to use shimmying. At the peak of the tick feeding period in late winter, moose may spend about 2.5 hours per day grooming, but infested moose also have friends that help – magpies, ravens, and gray jays perch on moose and eat the winter ticks, similar to oxpeckers that feed on ticks of zebras, giraffes, and other African mammals. Birds will also forage for engorged ticks on the forest floor in the spring.

Grooming incurs a large energetic cost, which is particularly serious for pregnant cows or moose that are undernourished, and it is likely that infested animals are less vigilant and therefore more susceptible to predators. With extensive damage to winter coats, ticks could conceivably also induce hypothermia during cold periods, although ghost moose usually appear in March and April when cold is less severe, so hypothermia is probably less important than blood loss, decreased fat stores, microbial diseases, and the extra energy moose have to use to groom.

Ticks Are Trouble

There is no doubt that winter ticks are a significant factor affecting most individual moose. Could they also affect moose populations? Because moose are important big game animals, wildlife management experts spend much time and money on the difficult job of monitoring moose populations to determine sustainable hunting limits. In North

America, moose in some areas are doing well (perhaps too well), but in others they are declining.

Populations in the Canadian provinces of Saskatchewan (48,000 moose) and New Brunswick (31,000) seem to be increasing, as well as those in Cape Breton Island in Nova Scotia (but not the mainland); in Newfoundland, there is a superabundance. Historically, moose were not native to Newfoundland, but four moose from New Brunswick were introduced in May 1904 (when they would be tick-free) to colonize the island. With no predators, no winter ticks, and an abundance of food, the population exploded and is now about 115,000 and growing. This increase in numbers is now a problem because moose-automobile collisions are becoming common, leading to great damage to vehicles, injuries, and deaths.

However, moose populations in many other regions are declining drastically, by 50 percent or more. Indigenous people, hunters, trappers, and wildlife biologists are seeing alarming decreases in moose populations, stimulating governments in Manitoba, Ontario, Nova Scotia, and British Columbia to introduce management plans. Are winter ticks causing these declines?

Making a Bad Situation Worse

In south and western Quebec, where moose numbers are shrinking, winter ticks infest more than 90 percent of the population, and ticks are now more common in many areas of the southern part of the moose range, including Minnesota,

Michigan, and parts of New England, than they were previously. Ghost moose sightings are increasing, and the likely culprit is climate change.

Warmer weather and the success of winter ticks are tied together. Warm, humid conditions in late winter and early spring promote increased survival of egg-filled female ticks in the leaf litter; warm, sunny, humid weather in summer allows more tick eggs to hatch and more seed ticks to survive; and mild, dry, autumns facilitate transmission of ticks to moose and lead to more severe infestations. Although moose are superbly adapted to thrive in extreme cold, they come under physiological stress when temperatures rise above 20°C (68°F). This combination of factors, plus others such as habitat loss, competition with white-tailed deer, and other infections we've already seen, such as liver flukes (chapter 4) and brain worms (chapter 10), are likely all contributing to the decline of moose.

Even more complex factors may also be playing a role. As we saw in chapter 10, a warmer climate has led to more frequent outbreaks of forest pests like mountain pine beetles. In British Columbia so many trees have died due to the beetles that logging activity has broadened to include salvaging dead trees. Loss of habitat through pine beetle infestation – something driven by climate change – creates a negative feedback loop where the beetle infestations also contribute to more frequent and intense wildfires, which are themselves exacerbated by a warming climate. Logging roads, cut lines, electric power lines, and oil and gas infrastructure break up old-growth boreal forests and encourage an increase of shrubby understory vegetation, which may be ideal food for moose and deer but also creates "highways" for predators like wolves. Winter tick–infested moose are likely weaker and less predator vigilant, so in this subtle way, pine beetles may also be affecting ticks, moose, and wolves. The specific impact of increasing wildfires on the moose–tick relationship is currently unclear, but what is clear is that ecosystems like the boreal forests of North America are changing fast and suffering as the planet warms.

One clear lesson that ghost moose can teach us is that the effects of climate change will have complex, interrelated, and unpredictable consequences. Ghost moose have likely been a natural part of the ecological interplay in northern forests for thousands of years, and moose die-offs due to ticks have probably occurred in the past but were likely self-regulating – as moose populations declined, there would be fewer hosts, so winter tick populations would decline too. This natural balance, however, appears to have now been upset. With an ever-warmer world, moose ticks will likely expand their range north and contribute to the continuing decline of these amazing mammals, perhaps, sadly, even to their extinction.

16
Sex and the Single Schistosome

At Tiputini Biodiversity Station, one of the biologically richest places on Earth, in the lowland tropical rainforest of eastern Ecuador, I hiked to a small, unobtrusive ravine that had a wet, mucky bottom, heavily marked with animal tracks. Even though the site was superficially nondescript, the researchers at the field station had set up a video camera trap there because this magic place attracts all kinds of wildlife – from jaguars, peccaries, and caiman to parrots, spider monkeys, white-tailed deer, and short-eared dogs, among many others. The attraction there was salt, a very rare commodity in tropical rainforests, as you may remember from our encounter with sloths and moths in chapter 13.

In the same way, waterholes (predictable sources of water) in Africa are wildlife magnets attracting all kinds of animals from zebras and wildebeests to elephants and baboons. The concentration of prey attracts lions and other predators, so like the salt lick area in the rainforest, these localized sites become a focus of biological interactions such as predation. These sites are also hot spots for the transmission of parasites and diseases. For instance, the deadly disease anthrax (caused by *Bacillus anthracis* bacteria) is spread among many animals that use waterholes. Similarly, a weird animal parasite, unlike most other animals of its type, gets spread into many mammal hosts (including humans) at waterholes. These unique parasites are blood flukes.

The Weird World of Blood Flukes

Blood flukes are schistosomes – parasitic flatworms (phylum Platyhelminthes, class Trematoda) that live in the blood vessels of birds and mammals. Their name "schistosome" means "split body" and refers to male worms that have a groove, called the gynecophoric canal, which runs the length of the male's body. Within the canal, female worms are held in a permanent copulatory embrace – schistosomes are the world's most romantic parasites (Fig. 17).

Besides the groove in males, blood flukes have many other unusual features. Most flukes are hermaphroditic with functional ovaries and testes in one animal, so by having separate sexes, which are different in shape and size, schistosomes are very unusual. The closest relatives of schistosomes are flukes called Spirorchids, which live in the hearts and arteries of turtles, and other flukes called Sanguinicolids, which live in marine fish, but these relatives are hermaphrodites like most flukes.

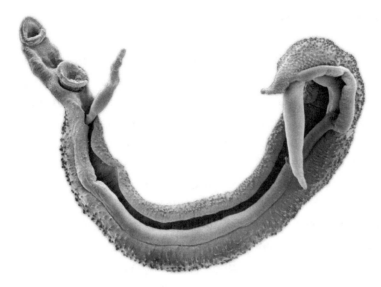

Figure 17. An adult male and adult female blood fluke, *Schistosoma* sp. The female is held in the gynecophoric canal of the male. Source: The Natural History Museum / Alamy Stock Photo.

Another unusual feature of blood flukes is that they are not picky about their vertebrate hosts. They infect birds, mammals (including rodents, primates, carnivores, even-toed hoofed mammals such as giraffes, and odd-toed ones like zebras), and for one species, freshwater crocodiles. Another odd characteristic of schistosomes is that they live in a unique habitat – they reside in blood vessels that drain their host's intestines and carry nutrients to the liver and also in veins near the urinary bladder. These unusual habitats of blood flukes may explain why schistosomes have two differently shaped and unequally sized separate sexes and also why they have strange courtship, unique mating behavior, and kinky sex.

An Unusual Home

In vertebrates, the hepatic portal part of the circulatory system includes blood vessels that occur between two capillary beds; one capillary bed is close to the intestine, and the other is in the liver. Digested nutrients from host food in the gut are absorbed by intestinal capillaries and taken by veins to the liver, where nutrients are absorbed, transformed, and stored by liver cells. Because schistosomes live in blood vessels, they can feed on blood cells, but the worms also get the first pick of their host's digested nutrients – simple sugars, amino acids, some lipids, vitamins, and minerals. Another benefit of living in their neighborhood is that it has been exploited by few other parasites, so competition for space and resources is unlikely. As a result, living in these portal blood vessels means there is plenty of rich, nutritious food for the worms.

However, living in this food-rich site also comes with challenges. First, worms are exposed to the full force of their host's immune system. Worm secretory and excretory products and eggs provoke host immune responses, and the host's protective cells and chemicals are constantly bathing the worms. Infected hosts produce antibodies in the blood, which can bind onto the body surface of worms, mark the parasites as foreign tissue, and recruit and activate host white blood cells that try to break down the worms. Antibodies can also attract and amplify complement blood proteins – chemicals that puncture blood fluke cell membranes. So, by living in the bloodstream, schistosomes are constantly bathed by chemicals and cells that are there to kill them.

Worm Tricks

Consequently, blood flukes have evolved tricks to avoid these host defenses. For instance, the outside surface of blood flukes, like other flukes and tapeworms, is composed of live cells (the surface of our and most animal skin is composed of dead cells, filled with keratin). However, unlike most fluke cells, these have not one but two cell membranes. Thus, when one surface of a blood fluke is attacked and covered with antibodies, it can simply be sloughed off, exposing a chemically different surface to the host. The host then has to recognize the strange tissue and mount a new attack; the worms have changed clothes to confuse their host. More remarkably, schistosomes can also camouflage themselves by coating their surfaces with host molecules that mimic the host's tissues, thus hiding from the immune system. In addition, blood flukes can suppress their host's immune system so that the attack on them is not as strong as it might be.

Shockingly, it has also been discovered that schistosomes not only try to evade host defenses but in fact need the host's immune response in order to stimulate their growth and egg production (host chemicals usually secreted by white blood cells are recognized and used by the worms as developmental cues). Adult blood flukes have not only had to evolve many counter-measures to survive in an unfriendly, hostile habitat but have taken them one step further and use host defensive molecules as fertility drugs.

Staying Put and Getting Out

Another challenge for blood flukes because of where they live is holding on and resisting blood flow, which is sweeping them toward the host's liver. This task is particularly important for eggs, as blood flukes have to get their eggs from the bloodstream into the outside world in order to get their offspring into another host. The parasites try to deposit their eggs on the inner lining of small veins, as close to the intestine as possible. To assist, schistosome eggs have a small spine that catches onto the inside lining of a blood vessel. The eggs then work their way through the blood vessel wall and the intestinal lining so that they can pop out into the intestinal canal, where they will be expelled from the host when it defecates (or, for those that live near the urinary bladder, urinates). To accomplish this task, blood fluke eggs utilize the host's intestinal muscle contractions and also release cell-breaking enzymes. In addition, it is likely that schistosomes use their host's immune reactions to assist their eggs in exiting from a host; for instance, eggs release antigenic chemicals, which provoke an immune response and affect their host's intestinal muscle contractions, resulting in more eggs getting passed.

Life Outside

When discharged in freshwater, eggs hatch and release small, ciliated non-feeding larvae called miracidia. They have sensory structures that help them find a suitable intermediate host, a freshwater snail. Although the adult flukes are not choosy about the vertebrate they infect, their larvae are very particular about the kind of snail they use. Once a miracidium contacts a suitable snail, it sheds its ciliated skin, penetrates through the snail's tissues, and becomes a sausage-like larva that can absorb nutrients, grow, and reproduce by cloning. These sausage-like stages are called sporocysts. When big enough, sporocysts also reproduce asexually, eventually forming larvae called cercariae, which we first came across in chapter 4. As with the giant liver flukes in that chapter, one single blood fluke miracidium can generate hundreds to thousands of cercariae.

Cercariae are continuously shed into water from a live, infected snail. These larvae have a characteristic forked tail that they use in a thrashing, darting style of swimming – they are non-feeding, free-living, and use chemical cues to search for the skin of their mammalian or avian host. When they find a vertebrate host, they will use muscular contractions and enzymes to penetrate its skin; they are not very selective – for instance, duck schistosomes will try to penetrate mammals (including humans). This attempt is usually unsuccessful, but the dead cercariae stuck in the skin do cause an itchy dermatitis in the unfortunate person who was swimming in the snail habitat water – swimmer's itch.

Schistosome Dating Sites

At the skin of a usable vertebrate host, the tail of the larva drops off, and the head of the cercaria forms a stage called a schistosomule. It moves from the skin into the circulatory system and travels to the heart, lungs, and eventually into the liver. By about 15 days after penetrating the skin, the 1 cm (0.4") long worms are now sub-adult males and females. The host liver is like a blood fluke singles bar – male worms secrete pheromones that attract females, and it's in the host's liver that pairing occurs. Hooking up is not random – females will reject worms that are not the same species, and when several potential mates are present, they will pair with the larger, more muscular males. Schistosomes usually have biased sex ratios, with more males than females; males are thus constantly on the prowl for females. There is sexual competition among males, and sometimes mates are switched as male worms can displace other males from pairs.

For schistosomes, this sexual dance in the host's liver is very important. Female flukes cannot complete their growth and mature sexually unless they are held in the gynecophoric canal of a male, but once paired, worms can reach their full adult size (about 1.5 cm or 0.6" long) and mate. However, the most important function of the sexually dimorphic worms pairing in the host liver is yet to come.

Going Home

After pairing, with the female firmly embedded in the body canal of her mate, the newlyweds begin a long migration from tiny blood vessels in the liver, upstream, against blood flow, to the smallest hepatic portal veins of the host's intestine. Some species migrate even further to blood vessels near the urinary bladder of their hosts. The outside back surfaces of muscular males are covered with gripping bumps and small spines, and by using muscular suckers around their mouths and on their ventral surfaces, as well as body contractions, the pair crawl along like caterpillars until they reach their final habitat – the smallest blood vessels near the intestine or bladder that they can wedge themselves into. Here, a mated female worm can begin to produce eggs and use her slender, lithe body to deposit them on the inside lining of the fine blood vessels. Without a male partner, an unpaired female fluke cannot successfully complete the migration from the liver, as only the male has the strength and appendages necessary to complete the journey.

Besides carrying their brides home, muscular males also help their mates feed – they pump host blood into their partner's digestive tract. While she is held in a permanent embrace within a male's body canal, a willowy female fluke can lay about 300 eggs per day and live for years (in one human case, worms lived for 28 years!). Lab studies show that widowed females (females that were separated from their partner) can survive for months but not produce eggs.[1]

How Blood Flukes Cause Trouble

Despite the painstaking efforts a schistosome couple uses to deposit eggs in the portal vessels, about half of the eggs still get swept away to the liver. Also, many eggs that start to work their way through the blood vessels and intestine never make it. Backswept eggs in the liver and those trapped in intestinal tissues are fiercely attacked by host defense responses, resulting in inflammatory reactions that create stringy nodules called granulomas. The role of granulomas is controversial – some think they protect hosts by walling off fluke eggs and protecting host tissues that are near eggs from damage, but other evidence suggests that they may also facilitate the passage of some eggs into the host's intestine. If the latter is true, blood flukes may again be manipulating host defensive reactions to their benefit.

The "smoldering" inflammatory reactions and blood loss in an infected host eventually cause circulatory problems, as well as enlarged spleens and livers, gastric varicose veins, anemia, lethargy, and other problems. Over years, normal intestinal and liver tissue will be replaced by non-functional scar tissue and tumors. The disease schistosomiasis is a chronic, debilitating, long-standing health problem that can lead to the death of the hosts (ecologically, it could be one factor that helps predators to hunt and kill older prey animals).

One reason blood flukes do not kill their hosts more quickly is because hosts do not get superinfected with many pairs of adult worms. Once established in the hepatic portal system, blood flukes stimulate their host to produce an effective, protective immune defense – not against them but against incoming schistosomule larval stages. This kind of host defense, which prevents superinfections, is called concomitant immunity. It benefits hosts by controlling the parasite population but also benefits the worms because it reduces competition for habitat space and helps to keep the host alive for a long time so that the reproductive output of the parasites is maximized. Long-lived adult worms are constantly vaccinating their host against larval worms and preventing large populations of flukes from infecting one host.

The Waterhole as a Hotspot of Infection

All these remarkably specialized adaptations that mammalian blood flukes have suggest that their evolution would demonstrate a long history of host specificity and cospeciation. That is not the case, however, at least for mammals. Blood flukes occur in crocodiles (one species), birds, and mammals, but evidence suggests that the worms evolved first in crocodiles (or another kind of cold-blooded aquatic host). Only later were water birds such as ducks captured, and then mammals. Mammals using waterholes would be concentrated in the same habitat as other potential hosts, and if emergent vegetation like rushes and reeds were present, along with snails, then blood fluke cercariae would be common and would attempt skin penetration into any animal trying to take a drink or to cool off there.

Evolution of Blood Flukes

Examining the diversity of blood flukes and determining their evolutionary history gives some fascinating ideas about how they came to have separate sexes, why the two sexes are so different in body shape, and how mating strategies developed.

In male worms, different kinds of blood flukes vary in the extent of the size of their body canal – from being almost the whole length of their body to the canal being absent – and the number of testes vary from 1 to more than 500. Research has shown that there is a trade-off between these two characters (size of canal and number of testes). Males with more muscles have better developed gynecophoric canals but fewer testes, which suggests that more muscular flukes are monogamous, pairing with one female and staying with her for life, while weaker flukes with many testes are promiscuous and try to fertilize many females. When biologists worked out an evolutionary family tree of blood flukes, they noted that worms started as muscular types, with longer canals and fewer testes. The flukes that evolved to infect birds are weaklings and have secondarily become promiscuous. Mammalian hosts tend to have high blood pressures, so muscular males have a better chance to complete the upstream liver-to-intestine migration, and having separate sized sexes was selected for by the need for females to be delicate enough to fit inside males and to deposit eggs on the lining of the smallest blood vessels.

Modern History of Blood Flukes

According to the World Health Organization, human blood flukes caused misery for about 251.4 million people in 2021, and transmission was reported from 78 countries.[2] Six different species of blood flukes, all in the genus *Schistosoma*, infect humans in Asia, South America, and Africa.

People living in poor and rural communities that depend on agriculture and fishing are most at risk, but increasing numbers of tourists are also getting infected. Molecular genetic studies support the idea that human schistosomes originated and diversified in Asia, and human blood flukes likely came from worms of herbivorous mammals.[3] Human worms invaded Africa using migrating mammals about 12 to 19 million years ago (mya) and diversified into the main types that infect people today between 1 and 4 mya. In addition to all the other horrors of the slave trade, African people infected with schistosomes carried the disease to South America and the Caribbean.

Besides humans and other primates, blood flukes occur in many mammals, including rodents, odd-toed grazers like zebras and rhinos, and even-toed herbivores such as hippos, giraffes, peccaries, and antelope. Even elephants with their tough skins can be infected by blood flukes. Although schistosomes have many specializations, including different-shaped sexes, their natural history indicates that they have colonized a variety of vertebrate hosts surprisingly well. When animals share still or stagnant water surrounded by stands of emergent vegetation supporting snails, like waterholes, schistosomes are fruitful and multiply.

The Benefits and Costs of Sex

According to biologists, the meaning of life is reproduction, but the pressures and constraints faced by parasitic blood flukes to successfully reproduce are many. Unlike most animals of their kind, they have evolved separate sexes that are shaped very differently, so their biology requires that their hosts be infected by at least two individuals of opposite sex. Most parasite populations are "over dispersed," meaning that most hosts have few parasites, while a few hosts have large infections. But due to concomitant immunity, blood flukes in host populations are more evenly distributed. As a result, most hosts in a population are infected by only a few pairs of worms, but hosts and their worms survive for years. Consequently, in host livers there must be strong sexual selection – females choose hunky males, which reinforces the physical differences between male and female blood flukes in mammals.

The two blood fluke sexes have very different functional roles to ensure that offspring can be transmitted successfully: males must be strong and muscular to migrate and hold the mated couple in place, while females must be thin and delicate to deposit eggs where they have the best chance of getting out of the host. Besides being necessary to get to and live in their host's circulatory system, separate sexes help schistosomes to increase genetic diversity, which is important in ensuring that offspring are variable and can respond to key changes in the worm's world such as variations in host defenses. This genetic variation has contributed to blood flukes' exquisite ability to handle and exploit host immune systems and perhaps to their ability to invade many different kinds of final hosts. Because many human health conditions, such as celiac disease and multiple sclerosis, may be due to inappropriate immune system reactions, studying blood flukes may show us new ways to try to adjust immunity and to control these problems.

Unfortunately for many people living in tropical areas, the meaning of life for schistosomes (sexual reproduction) negatively impacts quality of life for their hosts. Today, migrations and displacements of people are introducing schistosomiasis to new areas, and human changes to ecosystems and watersheds, including the construction of hydroelectric dams and the reservoirs that form behind them, increase blood fluke infections if they are present. Growing levels of poverty are increasing the prevalence of the disease, and if global climate change causes warmer, wetter conditions, then sadly, we are likely to see more human suffering due to these remarkably adapted parasites.

However, the unique sexual behavior that has made schistosomes so successful may become their Achilles' heel. When gene transcripts (RNA copies of genes) from ovaries and testes of paired and unpaired worms were examined, geneticists found more than 7000 types of molecules, but about half of these (3843, including 3600 in ovaries and 243 in testes) were made only when worms occurred in pairs, including those affecting activation of genes involved with worm stem cells and nervous system functions. These genes may give us valuable clues for developing hundreds of new drugs that would interrupt schistosome maturation and egg production. Because blood flukes have to be sexually dimorphic (have different body shapes and sizes) to successfully infect hosts, we can use the genes that control this feature to eradicate them.

Today, there is little hope of developing a blood fluke vaccine, and only one drug (praziquantel) is used for treatment, which in itself is very concerning since these resilient and adaptable worms may soon develop drug resistance. Having more than 200 new possible targets for drugs that would stop schistosomes from making their deadly eggs gives new hope for stopping this terrible disease. Studies using human volunteers who agree to be infected with only male *Schistosoma mansoni* are now underway.[4]

Blood flukes are remarkable and unique parasites that have evolved to infect many mammals and have developed amazing strategies to survive in an inhospitable environment. Perhaps soon we will be able to take advantage of their special biology, not only to control them but to treat other diseases too.

17

The Trickster: Coyotes and Their Parasites

As our population continually expands, more and more of us are living in cities. To accommodate suburbs, cities voraciously sprawl into rural and natural areas, which brings us increasingly into contact with wildlife. Sometimes wild animals will try to move away from urban areas, but sometimes they adapt. For instance, songbirds will change the frequency of their songs to cope with traffic noise, and rodents will evolve ways of handling deadly toxins so they can eat the moldy garbage that we litter our cities with. One of the most adaptable animals becoming integrated into our urban wildlife are coyotes. Across most cities in North America, people are coming into conflict with coyotes – they scavenge our garbage, kill family pets, and even aggressively attack joggers in parks. But few of us are aware that coyotes are introducing a deadly parasite into our cities as well.

The Trickster

One of the rich cultural legacies of Indigenous peoples of North America is that they have collected natural history information for generations and passed it down through the telling of stories. Many cultures feature stories about the character Coyote, who is often given male anthropomorphic character traits while retaining physical features of the animal. Although the stories inevitably vary from culture to culture, Coyote is often portrayed as the Trickster – giving the gift of fire that he had stolen from the gods to people. Sometimes Coyote is conniving and uses lies to deceive people or impersonate another person, or even a god, to get what he wants. From these stories, we learn that Coyote is clever, cunning, and resourceful – a real survivor.

Coyotes (*Canis latrans*) are true North American natives, having evolved in the southern plains of North America about 800,000 years ago. The likely ancestor of coyotes was a small wolf (such as *Canis edwardii*), fossil remains of which occur in the southwestern United States and date back three million years. The environment of North America changed over time and selected for smaller, fleeter canids, some of which migrated to Eurasia, but *Canis latrans* stayed. While wolves (*Canis lupus*) hunting in packs were large enough to feed on the huge herds of large grazers in the Pleistocene, smaller, cunning coyotes took smaller prey, scavenged on the remains left by wolves, and ate lots of plant matter too.

When the prey base of wolves began to decline, wolf populations shrank, but coyotes thrived. They adapted by becoming smaller and thus needing fewer calories of food energy. But coyotes also had another trick – they developed a special kind of social life called fission-fusion, a social system in which there is great flexibility. Animals can group together in packs and be gregarious, or if circumstances warrant, they can be solitary. This flexibility allows coyotes the ability to colonize and take advantage of a variety of habitats, and it also gives them the means of surviving periods of time with food shortages or outbreaks of disease. Coyotes sometimes work together in packs to kill larger prey and to protect themselves from predators and competitors like wolves, but when necessary, they can be solitary hunters and take rodents, rabbits, and insects, or scavenge, or eat berries or other fruits. Coyotes have killing canine teeth but also have crushing and grinding molars, so they can follow an omnivorous diet. These varied characteristics permitted coyotes to survive, even after humans came to North America, and allow them to continue to thrive today.

Humans and Coyotes

Indigenous people have coexisted with coyotes because they are not large enough to be dangerous predators or serious threats to humans; nor are they competitors for game. In fact, both coyotes and humans have the same, flexible social system that made them resilient to changing conditions or circumstances. Both have the tools needed to exploit a variety of foods, and if worse came to worst, both could migrate to new habitats. It is no wonder that many stories of Coyote came about and that these stories had morals that taught people lessons for survival. In coyotes, people saw their reflection as clever, resourceful survivors.

When Europeans came to North America, they declared war on wolves. Folk tales and stories like *Little Red Riding Hood, Peter and the Wolf,* and *The Three Little Pigs* portrayed wolves as dangerous, devious predators. As settlers steadily moved west, converting forests and prairies to farms and ranches and keeping herds of domesticated animals, they came to detest wolves. Bounties were placed on wolves, and people were encouraged to kill them on sight. Agricultural societies, backed by state, provincial, and federal governments, paid for extermination programs using professional hunters and trappers. Strychnine-poisoned baits were used. These measures, as well as habitat loss and reductions in prey animals, almost completely eradicated wolves from much of North America. One unexpected outcome of this policy, however, was that, by reducing the population of a predator and competitor, coyote populations prospered.

After wolves were eliminated, ranchers believed they were still at risk of losing valuable animals to coyotes. Since government-sanctioned killing programs had been successful at extirpating wolves from most of the continental United States, people turned their attention to coyotes. Try as they might, the same efforts employed to eliminate wolves did not work, so new methods of killing coyotes were developed. In Montana, the strategy involved using a parasite – *Sarcoptes scabiei.*

A Human Parasite in Wildlife

Sarcoptes scabiei are itch mites. Mites are tiny arthropods related to scorpions, spiders, and ticks (see chapter 15), and they cause a skin disease called scabies in humans and mange in coyotes (Fig. 18). The mites are microscopically tiny, being only 0.2–0.5 mm (0.008–0.02") in length. They have globular bodies and look like little turtles, with rows of short spikes projecting from the striated cuticle on their backs. Mites, like ticks, have four pairs of legs, which bear bell-shaped sucker discs (caruncles) on the ends, and legs three and four terminate in long, hair-like projections called setae. The last section of their legs also have a couple of bladelike claws. If a female mite gets on the skin of a coyote, it uses its mouthparts to excavate a tunnel in the outer layer of the epidermis and then continues tunneling to the softer skin layers below. The mite's spikes irritate the tunnel walls in the skin and induce tissue fluid to ooze into the tunnel – this fluid, along with host cells, is consumed by the mites.

For four to six weeks, mites burrow, and each female lays three to four eggs per day in a tunnel. The eggs then hatch, and six-legged larvae emerge – some stay in their birth tunnel, while others break out onto the surface of the host's skin. The larvae molt twice to form eight-legged nymphs, which then molt again to form adults. Completion of the life cycle, from egg to adult, takes about two weeks. Mites are transmitted between coyotes by direct skin-to-skin contact, or if the humidity is high and the temperature cool, mites can live off a host for several weeks (for example, in a den) and infect other coyotes indirectly.

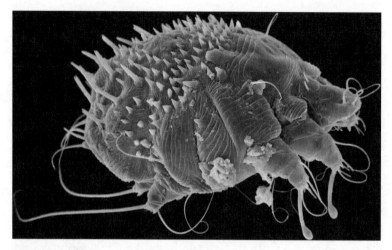

Figure 18. The mange mite, *Sarcoptes scabiei*. Source: The Natural History Museum / Alamy Stock Photo.

Sarcoptic mite burrowing and mite feces in the tunnels cause intense itching, red inflamed skin eruptions, papules, sebaceous gland secretions, and fur loss. Scabby crusts form on the hide, and coyotes take on a foul smell. Hair is lost on the legs, the tail, the ears, the muzzle, and the torso so that, in an advanced case, 80 percent of the hair coat is gone (Fig. 19). One mangy coyote even chewed off the last vertebrae of its tail – the bones were found in its stomach. Before death, mangy coyotes become listless and emaciated, and lose their natural fear of people – the coyotes you see in your yard often have mange. Sarcoptic mange causes a prolonged (about 60 days), agonizing, horrible death, and at death the population of mites on a coyote ranges from hundreds to several thousand per cm^2 ($0.4"^2$) of hide, which would result in about three million mites per animal. Because it takes so

Figure 19. A coyote with an advanced case of sarcoptic mange. Source: Cameron Rognan.

long to progress, the infection can be readily spread from adults to pups or between siblings and mates.

In 1905, the Montana State Legislature passed a law requiring veterinarians to intentionally introduce sarcoptic mange into those "parasites of civilization," coyotes. The program was successful in one sense because today it is common to see mangy coyotes. However, if the reason for using this biological weapon was to exterminate coyotes, it was a dismal failure. Today, coyotes are more abundant than ever and seem to be expanding their range. They have become successful cosmopolitan mammals and are often seen in cities ranging from Los Angeles to Boston and Miami to Edmonton.

Coyotes Come to Town

Are urban coyotes different from their wild cousins? Besides mange, what diseases and infections do they carry, and might some of these be passed to us or our pets? Are urban coyotes as healthy as wild ones? Surprisingly, few researchers have looked at the parasites of urban coyotes to answer these questions. One study in Tucson, Arizona, found that urban coyotes were like wild coyotes with regards to bacterial and viral infections and that these infections were likely to cause death only during stressful times, such as food scarcity, a high population density, or if parasites became common.[1] The researchers examined 19 coyotes, and although about 14 percent of them had mange, they did not report finding any internal parasites.

In Calgary, Alberta, a city with a population of about 1.1 million people and an unknown number of urban coyotes, 61 hunted and road-killed coyotes were examined for parasites.[2] The animals were found to be infected with seven different species of parasites, the commonest of which was a roundworm, *Toxascaris leonina*, a cosmopolitan parasite of canids that has a direct life cycle – eggs on the soil (which are very resistant to freezing) are ingested. After undergoing a migration in their host that involves molting, adult worms reside in a coyote's intestine, where they cause little damage unless infections become very large (100 worms or more), in which case they can cause intestinal blockages. If humans accidentally ingest eggs, the juvenile worms may cause damage as they wander around inside the wrong host – a condition called visceral larva migrans that we saw can also arise from infection of *Baylisascaris* from raccoon feces (see chapter 12).

The second most common parasite in urban coyotes in Calgary was another roundworm called hookworm (*Uncinaria stenocephala*). Hookworms also directly infect their hosts, but in this case, small juveniles that have hatched from thin-shelled eggs in moist soils penetrate the paw pads of coyotes. After migrating through the lungs and molting, adult worms infect the intestine, where they live by sucking blood. In large infections of more than 100 worms, they can cause anemia, especially in pups. Transmission likely occurs in damp, sandy soil near dens. The low number found in Calgary coyotes suggests that hookworms are not important in causing disease, but they may be a concern because domestic dogs can get infected. There are also a few reports of human infections.

A Dangerous Disease

Another parasite that was detected in Calgary coyotes, however, may be very important – not to coyotes but to people. Researchers found that 30 percent of coyotes in the city were infected with the world's smallest tapeworm, *Echinococcus multilocularis*. If you recall from chapter 11, tapeworm bodies are usually composed of hundreds of repeated body segments (proglottids), but *E. multilocularis* is made up of only three or four. The first segment (the scolex) attaches a worm to the coyote's intestine. Behind the scolex is a segment with female and male reproductive organs, while the last section has a uterus filled with embryonated eggs. The eggs are passed out of the coyote onto the ground.

At this point, dangerous trouble begins. If any warm-blooded animal (rodents, rabbits, birds, humans) ingests the eggs, they hatch in the host's intestine, releasing small, six-hooked larvae. These larvae penetrate the intestinal lining, get into the bloodstream, and are distributed to various internal organs such as the liver and lungs. There, they metamorphose into multichambered, fluid-filled cysts called hydatids, the same cysts that caused the cow moose in the introduction to be caught by wolves. The thin-walled hydatid cyst infiltrates host tissues, buds off chambers, and grows like cancer. The inside lining of a hydatid is germinal tissue, which generates protoscolices (little tapeworm scolices) called hydatid sand. If the infected intermediate host is eaten by a coyote or other dog, the protoscolices will each develop into a little adult tapeworm in the canid's intestine.

It is easy to imagine that, if urban coyotes are infected with *E. multilocularis*, domestic dogs running around in off-leash parks may find a dead squirrel, mouse, or vole that has hydatid cysts. After eating it, our best friend may host adult tapeworms and be passing thousands of worm eggs per day, perhaps in our backyard or neighborhood playground. The potential for a dangerous human infection is great, and indeed we are unfortunately now seeing human cases of hydatid disease in Alberta.

Because of this danger, and to see if the parasite community of Calgary coyotes is typical or otherwise, a study of 14 urban coyotes was conducted in Edmonton, Alberta, a city of about 900,000 located 300 km (186 miles) north of

Calgary.[3] All coyotes examined were infected by parasites, with an average of 4 kinds per host, and even with a small sample size, there were 10 different animal parasites, almost double the number found in the larger Calgary sample. Surprisingly, the commonest parasite found was the dangerous tapeworm *Echinococcus multilocularis*, with an average population of more than 4700 worms per coyote (one was infected by more than 16,000 worms). Such large infections in coyotes are not surprising because an animal needs only to eat one cyst in an infected rodent to acquire hundreds to thousands of protoscolices. Although massive, the infections in coyotes are not normally damaging to them – large populations of the tiny tapeworms are not an energy drain, and their tiny scolices seem not to damage their host's intestinal lining. What is surprising, though, is that urban coyotes are infected with *E. multilocularis* at all. Where did the infection come from?

Parasite surveys of canids in North America and Europe indicate that foxes are the normal host for *E. multilocularis*, and it appears they have only recently infected coyotes.[4] In the 1960s, parasitologists surveyed wild coyotes and wolves throughout Alberta and did not find these little tapeworms.[5] However, both wolves and coyotes were infected by a related species, *E. canadensis*, which, as we saw in the introduction, uses cervids (especially moose) as an intermediate host rather than mice and voles. Unlike dangerous *E. multilocularis*, this worm causes individual cysts in the organs of intermediate hosts and does not proliferate like cancer. It mostly uses wolves as a final host, although dogs that scavenge on or are fed moose and deer viscera also get infected. Although parasite surveys of wild coyotes were conducted periodically throughout the western plains of North America, it was not until 1978 that *E. multilocularis* was found in coyotes from southern Manitoba.[6] Why do we see this dangerous parasite appearing in urban coyotes only relatively recently?

Red foxes have a complicated history of colonizing North America. They probably immigrated from Asia across the Bering land bridge at the end of the Pleistocene, and they originally lived in the boreal and western montane areas but have extended their range southwards. Arctic foxes (*Vulpes lagopus*) are infected with a very dangerous European strain of *E. multilocularis*, but red foxes found in more southern regions of North America were host to a less damaging strain. Several times, however, European foxes have been intentionally brought to North America (as early as colonial times) to be hunted and for fur farms. It is likely that a European strain of *E. multilocularis* was accidentally introduced to North American red foxes with these imported animals.

Perhaps coyotes, which were rarely infected with the North American great plains strain of the little tapeworms, have become infected with the European strain due to habitat overlaps with foxes, and now the parasites are much more common. To make matters even more complex, the European strain of *E. multilocularis* is much more damaging in humans who get infected than the native North American plains type, so it's more of a worry. Also, a molecular analysis indicates that both red foxes and coyotes in Alberta may be coinfected with *E. multilocularis* and *E. canadensis*. In our study, all worms that we found in coyotes in Edmonton had the physical features of *E. multilocularis* and were gravid (filled with eggs), but we do not know what strain the parasites are, European or North American – researchers at the University of Alberta are determining the strain now.[7] If given the opportunity to scavenge road-killed deer and moose, urban coyotes can become coinfected with *E. canadensis*.

Are you confused yet? Well, the situation may be even more complex. The study of rodents living around Edmonton done in the early 1970s showed that about 50 percent of white-footed deer mice (*Peromyscus maniculatus*) had multilocular hydatid cysts.[8] However, no red-backed voles (*Clethrionomys gapperi*), which are common coyote prey, were infected. Urbanization tends to result in population increases of house mice (*Mus musculus*) and declines of deer mice. House mice are better intermediate hosts for *E. multilocularis* than deer mice. Urban sprawl often results in forest habitat being replaced by grassland (lawn) habitat, which also changes rodent communities in cities. Might the high prevalence of *E. multilocularis* in urban coyotes reflect changes in rodent populations in cities?

All these features of coyote infection with *Echinococcus* are mysterious and puzzling – I guess exactly what we should expect from the Trickster. Currently, not many human cases of hydatid disease have been reported in Alberta, especially in cities that have many urban coyotes.[9] However, the disease can be misdiagnosed as cancer, and because it is a slow-developing, non-reportable disease, more cases may occur than we are aware of. Because of the potential danger to public health that hydatid disease represents, it will be important to get a better understanding of the risks that urban coyotes pose.

Urban Coyotes Are Sick

Besides *Echinococcus* tapeworms, other parasites of urban coyotes and other evidence suggest that urban animals are less healthy than wild ones. For instance, most urban coyotes carry the roundworm *Toxascaris leonina* and two species of large tapeworms, *Taenia pisiformis* and *T. serialis*, in their intestines, which indicates that rabbits make up a significant proportion of their diet. Hookworms (*U. stenocephala*) occur in about a third of urban coyotes, and the coyotes had small numbers of lungworms and stomach worms too (*Oslerus* and *Physaloptera*). In the intestines of Edmonton coyotes, we also found two different kinds of flukes (parasitic flatworms) – *Alaria arisaemoides* and *Alaria americana*.[10] The flukes are not likely causing damage to their hosts but are interesting because they tell us something about coyote habitat use and diet. The flukes are common and widespread in wild coyotes and require two intermediate hosts in their life cycle – aquatic snails and tadpoles or adult frogs. Edmonton city coyotes must frequently forage in wetlands, near ponds and drainage control lakes – these areas may be hot spots for transmission of infections.

Besides the greater diversity and higher populations of these parasites, other indicators of poor urban coyote health included sarcoptic mange and lice (*Linognathus setosus*), which are rarely found on coyotes but occur more commonly on domestic dogs. We also saw cases of oral papilloma virus, which causes oral warts (small benign tumors around the mouth), indicating a host with a weak immune system.

Although urban coyotes are sicker than their wild cousins, the benefit of city living may be more food – urban coyotes are omnivores and take advantage of anthropogenic foods. The commonest food items found in urban coyotes are plant material, particularly berries from ornamental trees like mountain ash and chokecherries. Apples (especially crab apples), grass, and spruce needles are common too. Besides nutrients, there may be another reason that urban coyotes eat grass and berries. We noticed that the average number of parasitic worms in coyotes that had recently eaten plant material was a third of the number in coyotes that had no plant matter in their stomachs. Perhaps coyotes eating plants are not exposed to as many parasites, or urban coyotes may be self-medicating (as is seen in domestic dogs) by consuming high-fiber, astringent anti-parasite compounds.

Animal remains, especially of voles and hares and rabbits, were the second commonest food in urban coyote stomachs. These prey animals are intermediate hosts for many of the parasites we found. Finally, anthropogenic items such as string, rubber, and plastics made up about a tenth of coyote stomach contents. Recently, a wildlife biologist in Alberta was sent a coyote fecal sample because a city park warden who found it thought there was a large "worm" in it – this item turned out to be a condom!

Because of increasing public concerns about interactions with coyotes in Edmonton, researchers established the

Edmonton Urban Coyote Project to learn how many coyotes are present and to record the parts of the city that are most frequented.[11] Besides providing information that might mitigate human–wildlife conflicts, they learned a lot about food habits and health of urban coyotes. One surprising element that came from their research was that municipal compost sites attract coyotes. Using camera traps, they discovered that coyotes were eight times more likely to visit compost sites (versus natural areas like parks) and that the coyotes that frequented compost areas were almost five times more likely to have mange and ten times more likely to be infected with tapeworms. In addition, another problem for urban coyotes was identified – the compost sites had high concentrations of mycotoxins. Mycotoxins are poisonous chemicals produced by fungi (molds) that can cause acute effects, like vomiting, and chronic effects, such as immune system suppression and organ failure. For coyotes, living in cities is a trade-off – they get access to plenty of varied food resources, but they run the risks of becoming roadkill and of getting infections, diseases, and poisoned.

The Big Picture: What about Where I Live?

Biologists have been monitoring the parasites and diseases of coyotes in North America for more than 70 years now and have discovered some interesting trends. The assortment and types of parasites in coyotes tend to be predictably similar, regardless of whether the hosts are living in south Texas or northern Canada. However, the Trickster does take advantage of the abundance of local resources to survive – for instance, coyotes that live in areas with lots of lakes (like Minnesota) have more parasites that use fish as intermediate hosts.

Also, parasite communities in coyotes have changed over time – notably, *E. multilocularis* was absent in early studies but is now common. Urban coyotes have more parasites than wild ones, and each specific city may have its own characteristic coyote "parasite signature."

Species richness (the number of different kinds of parasites) and biodiversity (the variability and variety of parasites) are not directly related to latitude. Coyotes from warmer areas do not have more parasites than those from colder regions, but coyotes from drier locations tend to have richer, more diverse parasite faunas. Coyotes evolved on the arid grasslands of the central and southern plains of North America. Their dominant parasites (for example *Toxascaris leonina* and *Taenia pisiformis*) have a long coevolutionary history with them and are well adapted for transmission in dry, cold grassland habitats where small rodents and rabbits are abundant.

Today, coyotes are expanding their range, especially into wetter areas like eastern and southeastern North America (as far south as Panama). Currently, no postmortem surveys of coyote parasites from southern Ontario, Quebec, New England, New York, the maritime provinces of Canada, or the southern United States have been reported, but based on the trends we see, coyote parasite faunas may be very different (and likely less rich and diverse) than those from coyotes on the prairies and plains. Because parasites contribute to urban coyotes being less healthy, the impoverished parasite fauna of those in wetter climates may contribute to the success of coyotes in colonizing these areas. Biologists need to look at parasites of eastern coyotes, judge the health status of these animals, and gauge their potential to spread infections to humans – especially hydatid disease.

Living with Coyotes

Today, urban coyotes get in the news because of unfortunate interactions with us and our pets. Unlike the coexistence that occurred between Indigenous peoples and wild, sylvatic coyotes, urban coyotes occasionally come into our private spaces, raid our garbage, scavenge on fallen fruit, and kill our pets. For animals that are masters of adaptation, who have been able to handle trapping, hunting, poisoning, and attacks with bioweapons, we should not be surprised that coyotes can survive and even thrive in cities. Like it or not, coyotes are here to stay, and we have to learn to live with our new neighbors. To reduce conflicts, we can secure our garbage, pick up fallen fruit, keep our cats indoors, control dogs in off-leash areas, and haze visiting coyotes away from our gardens. After gardening, we should thoroughly clean our hands and wash any produce we harvest. Physicians should be alerted to the signs and symptoms of hydatid disease and need to be on the lookout for it – it will be an increasingly important zoonotic disease. Based on its evolutionary history of readily adapting to changes, the Trickster will continue to thrive and to adjust to his parasites and diseases, so we'd better have the necessary knowledge to deal with nasty disease surprises that Coyote may yet have in store for us.

18
Fleas: The Inside Story

Ever since we learned that dinosaurs such as *Tyrannosaurus rex* and modern birds are related and that some dinosaurs had feathers, I have wondered if dinosaurs were (like birds) infested with feather lice. My hope is that, someday, a paleontologist will discover a beautifully preserved dinosaur feather with lice attached. It would be amazing to find out how lice have changed over millions of years and if the kinds of lice dinosaurs had were similar to the kinds of bird lice that we think are the most primitive. Although no dinosaur lice have been found so far, these kinds of questions about parasites of long-extinct hosts may be answered by a parasite of mammals. Fleas normally infest mammal fur, but one strange type burrows under the skin, and this flea may hold answers about the parasites of some long-dead hosts.

Fleas

For most people, the word "parasite" evokes thoughts of disgusting creatures, indicative of squalor and filth, and if pressed for an example, fleas frequently come to mind. I'm not sure why fleas are so often considered the "textbook" example of a parasite. Could it be because they are ectoparasites, not hidden from view, and many of us have been pestered with irritating bites from dog and cat fleas? Maybe there is a deeper reason from the time when rat fleas were blamed for spreading the terrifying Black Death and killing half the population of Europe (see chapter 3).

In history, art, and literature, fleas have long been tagged as classic examples of parasites. For instance, in 1665 the curator of the Royal Society in London, Robert Hooke, presented what would become the world's most famous scientific image of the seventeenth century – a detailed microscopic diagram of a human flea, *Pulex irritans*, in his book *Micrographia*. He described fleas as "adorn'd with a curiously polish'd suit of sable Armour" with "multitudes of sharp pinns, shap'd almost like Porcupine's Quills." Regarding fleas as parasites, Hooke wrote: "This little busie Creature" sucks "out the blood of an Animal, leaving the skin inflamed." Given the negative connotations of fleas spreading deadly diseases, sucking blood, and attacking us and our pets, it is a hard sell for a biologist to convince anyone that lowly fleas are in fact amazing animals with a rich suite of specializations and a long history with mammals.

Flea Specializations

Fleas are evolutionarily specialized insects from the order Siphonaptera (*siphon* – tube; *aptera* – wingless). While most insects have two pairs of wings, fleas have lost their wings completely (Fig. 20). Instead, for locomotion they use their massive (proportionally for an insect) hind legs to leap onto hosts. They can propel themselves into the air more than 50 times their body length, using an elastic, rubbery protein called resilin. For comparison, if we were like fleas, a six-foot (1.8 m) tall human would be able to leap 300 feet (91 m) into the air from a standing position! Of course, simple up-scaling like this example is not true because of allometry (changes in body proportions like height do not change linearly with size) – but nevertheless, fleas are amazing athletes.

Stimulated when they sense a passing host (using heat, carbon dioxide, vibrations, and for fleas with eyes, shadows), they launch themselves into a wild somersault, trying to latch onto fur using metatarsal claws like grappling hooks. The size and shapes of these claws are tailored for attachment to their hosts' fur. Once on a host, fleas are laterally compressed so they can easily maneuver through a dense forest of hair. Flea bodies are adorned with backward-facing, chitinous spikes and spines. Some form rows like picket fences at the bottom of the head (a genal comb) and on the first section of their thorax (a pronotal comb), which help fleas resist being removed from their host. The spaces between the spikes in a comb have evolved to become directly related to their host's hair diameter, so as a result, fleas can hold onto their normal host without easily being

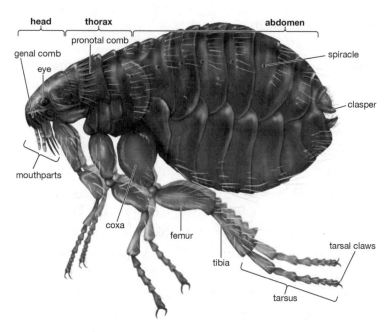

Figure 20. The anatomy of a typical flea. Source: Universal Images Group North America LLC / Alamy Stock Photo.

groomed. In addition, fleas are covered by plates of armor strengthened with internal rods, so they are very hard for hosts to kill by using sharp claws or nipping incisor teeth – crushing a flea takes a lot of time, skill, and effort.

Flea Diets

Fleas use their short, club-shaped antennae and little segmented appendages on their mouthparts to probe and taste host skin. All fleas are obligate blood feeders – before they can reproduce, they must feed using a blood-sucking tube that is unique among insects. Like most blood-feeding parasites, fleas inject chemicals from their salivary glands into their host, including anticoagulants that inhibit blood from clotting. Some of these chemicals stimulate an allergic response and are responsible for the itchy sore that develops at the feeding site. Unfortunately, spitting into host blood vessels can introduce many bacterial and viral diseases (such as the Black Death bacteria discussed in chapter 3). Fleas are wasteful feeders – semi-digested blood passes through them quickly and is squirted out of the digestive system into their host's den. This fecal blood in the den can later serve as food for larval fleas, and it also helps to keep the host den or nest more humid and cozy for flea larvae to survive without dying from desiccation. Besides feeding, adult fleas must find a mate in the dense pelage of their host.

Singing Fleas

To find mates in host fur, fleas "sing" by rubbing small spines on their hind legs against a serrated plate on the bottom of their thorax. Fleas create high-frequency, raspy songs that are detected by specialized sensory structures on the top surface of their abdomens, which look like satellite dishes (called sensillia). Once they have found a mate, males with their jaunty, upturned backsides work their way beneath the female and stimulate, hold on to, and support her using their antennae. In addition, males have a movable structure at the back of their abdomens, called a clasper, to grasp females. Complicated male reproductive structures (they superficially look like watch springs) are projected out and up to curl around the female's abdomen and inseminate her, and the sperm are then stored in a species-specific shaped structure, called the spermatheca, and can be used to fertilize eggs for days after mating. The sizes and shapes of the reproductive organs reduce the chances of reproduction between two different species of fleas, and for biologists, the structures also turn out to be useful features for identifying the insects.

After mating, female fleas eject white, oval eggs in the detritus at the bottom of their host's den but sometimes will lay eggs on host hairs, which fall off later. When eggs hatch, blind, worm-like larvae feed on fecal blood and other organic debris in the den, and some larvae may even cannibalize other flea larvae. They pupate in the den, and newly formed adults use a chitinous plate at the front of their heads to break open the pupal skin and emerge, at which point they are hungry and ready to find a suitable host and start feeding. Over evolutionary time, this life cycle has proved to be very successful, especially for infesting hosts that are habitual users of dens or nests.

Hosts: Animals That Live in Burrows

Today, there are about 2600 species of fleas, and more than 90 percent are parasites of mammals, especially rodents, although fleas are also found on insectivores such as shrews, marsupials such as opossums, bats, carnivores, rabbits and hares, and 1 percent of all species occur on armadillos, anteaters and sloths, hyraxes, and cloven-hoofed mammals, especially pigs. While most fleas have evolved to infest mammals – and a wide range of mammals at that – some species infest birds, and evolutionary biologists think that a host switch from mammals to birds happened at least four times in history. For example, rodent and marsupial fleas likely infested burrow-nesting birds like puffins and petrels, and squirrel fleas likely infested hole-nesting birds like owls. In the case of *Ceratophyllus lunatus* found on weasels, the infection likely went in the opposite direction, from birds back to mammals. Going back further in time, zoologists wonder where fleas, with their parasitic lifestyle and all the specializations they have, came from.

Flea History

Both molecular and anatomical evidence shows that fleas, which now are highly specialized as parasites, evolved from a common ancestor shared with little-known, free-living insects called snow scorpionflies (order Mecoptera, family Boreidae). Snow scorpionflies are small insects with beak-like mouthparts that are often found in moss and have adapted to live in cold conditions – they can be seen hopping about on snowbanks on sunny winter days, hence their name. Most telling about their relationship with fleas is that female snow scorpionflies are wingless, while the wings of males are small hard stubs used for grasping onto females during mating in a similar way to fleas. Loss (or reduction) of wings in snow scorpionflies could be an adaptation to reduce surface area for better survival in cold conditions – less surface area means they would lose heat more slowly.

Extinct fleas have been found in amber (fossilized tree resin) in the Baltic area and in the Dominican Republic. The Baltic specimens are placed in a genus that no longer exists, as no living types are known. However, the Dominican specimen is so much like living fleas that it has been placed in the large genus *Pulex*, of which there are many living representatives, including *Pulex irritans*, the commonest flea that infests humans. These preserved fleas date from 30 to 40 million years ago (mya), but because their features are identical to modern fleas, they do not provide any clues about how fleas evolved to become parasites.

There are Mesozoic fossils, however, which some scientists think are "pre-fleas" – if that is indeed the case, they may give us some clues about how fleas became parasites. In general, pre-fleas had heads with beaks, short club-shaped antennae, long legs, and well-developed claws, and they resemble perhaps the most frightening, creepy-looking living parasites of all time – bat flies (Nycteribiidae; Fig. 21). Bat flies are long-legged, spider-like ectoparasites that look as though they have no heads; in fact, their heads are folded into a groove on the back side of their thorax. One of the pre-flea fossils shows signs that pre-fleas were laterally compressed, as are all modern fleas.

There are two different ideas about the origin of today's fleas. Because of the resemblance between pre-fleas and nasty-looking bat flies, some researchers proposed that pre-fleas were adapted to infest the hair and bat-like

Figure 21. The creepiest-looking parasite – a "headless" bat fly (Nycteribiidae). Source: Gilles San Martin / CC BY-SA 2.0 DEED (https://creativecommons.org/licenses/by-sa/2.0/deed.en).

membranous wings of pterosaurs, extinct flying reptiles that lived during the Mesozoic. This hypothesis suggests that moss-living snow scorpionfly ancestors became pre-fleas that infested pterosaurs and later made the jump onto mammals – perhaps because their hosts became extinct. However, evolutionary biologists have pointed out that modern fleas specialized to live on bats today do not resemble pre-fleas at all, which suggests a second idea – that snow scorpionflies living in the moss lining mammal nests and dens were already wingless jumpers with mouthparts shaped well enough to allow for sucking. With frequent contact between these pre-fleas and mammals inside dens, the flea ancestors started to hop onto mammals and suck blood, and later became laterally flattened. This hypothesis removes the "middle-man" – pterosaurs – from the chain of events leading to modern fleas. Because it is a simpler explanation, most biologists think it is more likely. If the idea that scorpionfly ancestors living in mammal dens became fleas, which mammals could have been the original hosts?

First Hosts

Because rodents are the most commonly infested hosts of fleas today, logic would point to them as the first hosts, but this idea may be wrong. Fleas that have the fewest structural adaptations to be parasites (they lack combs and spines) are likely the most primitive kinds of fleas, and they infest mammals that most biologists also think are primitive – shrews, moles, opossums, wombats, and marsupials. So rather than rodents, perhaps these hosts were the first mammals to get flea-bitten. Detailed studies of molecular information – DNA and proteins of fleas, as well as analysis of flea physical adaptations – have shed more light on this question and led to some amazing suggestions.[1] First, they showed that fleas evolved long ago in the Cretaceous (35 million years before non-bird dinosaurs went extinct) in Gondwanaland (the supercontinent that included all of today's southern continents) at the time that the common ancestor of today's fleas started to diversify. However, fleas really spread and differentiated after the Cretaceous-Paleocene (K-Pg) boundary, 66 mya, when all non-bird dinosaurs went extinct. Animals that dwelled in underground burrows and dens (great habitat for fleas) made it through the cataclysmic asteroid event that wiped out the dinosaurs and host associations, and geographical distributions of fleas today strongly indicate that modern fleas survived in South America and then spread to become global parasites.

By matching the timing of flea evolution to host evolution, scientists concluded that the presumed Mesozoic "pre-flea" fossils did not lead to the evolution of fleas, so fleas were not parasites of dinosaurs or pterosaurs. Instead, strong evidence suggests that the basal group to all other fleas is a family called the Macropsyllidae. Today, these fleas are parasites of Australian marsupials. It therefore seems that fleas evolved not as parasites of dinosaurs or pterosaurs but as parasites of mammals, particularly early marsupials and placental mammals. Primitive egg-laying mammals (like the duck-billed platypus and spiny echidna) became hosts later, as did birds. So taken together, evidence suggests that fleas have always been blood-sucking parasites of mammals, but it also tells us that they have not always been restricted to living only on the outsides of their hosts.

Fleas That Live Inside Their Hosts

The great flea diversification after the K-Pg boundary mirrors the diversification of mammals, and yet even among such diversity, the twisted history of one of these groups of fleas is even weirder than most. Tungids (family Hectopsyllidae) are fleas that have very unusual bodies and a unique natural

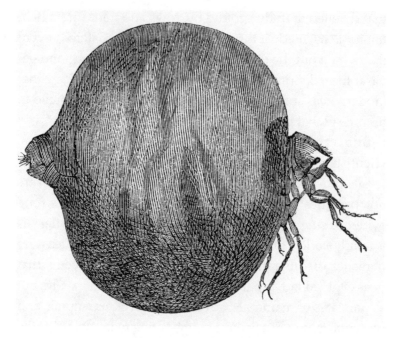

Figure 22. *Tunga penetrans* – neosome. A female sand flea with an extended abdomen. Source: FALKENSTEINFOTO / Alamy Stock Photo.

history, and as a result, they are the most specialized of all fleas. Tungids are stout fleas that have compressed thoracic segments and are almost completely devoid of spines – they are so weird that, when you see a female, it is hard to believe she is a flea (see Fig. 22). One species, *Tunga penetrans*, commonly called sand fleas, chigoes, or jiggers, is notorious because it is a generalist parasite that infects many hosts including humans, where it causes an infection called tungiasis. The disease occurs because sand fleas are not ectoparasites like all other fleas – they are endoparasites that live *inside* their hosts.

Female *Tunga penetrans* burrow into the skin of their hosts, often between the toes or – even more painfully – under toenails. Penetration triggers a host to encapsulate the flea, but the rear end of the female flea is placed at a hole in the capsule that allows her to be fertilized and also to expel eggs out of a host through the pore. Once trapped in a capsule, the female flea's abdomen enlarges more than 1000 times, forming a pea-sized, painful, ulcer-like sore. The globular female, now almost unrecognizable as a flea, is called a neosome (*neo* – new; *soma* – body). She feeds on host tissue fluids and expels hundreds of eggs onto sandy soil from the pore, where they later hatch. The spent female in the capsule dies and (best case scenario) is expelled from the host or is trapped in the capsule, where she rots. In the latter case, the capsule frequently becomes secondarily infected by bacteria, which can lead to gruesome infections including gangrene. The infection is so painful that there are reports that members of Christopher Columbus's crew, when they got infected in the new world, amputated their own toes.[2]

Today, sand fleas are found in tropical areas around the world. One human infection from Brazil reported 199 neosomes,[3] but dogs, pigs, rats, wildlife such as jaguars, and most domestic animals can also get infected. Because sand fleas live inside their hosts, they do not require specializations to grab onto fur or to avoid being groomed, so they can infect just about any mammal that spends most of their time on the ground. Although sand fleas evolved in sand soil habitats in the Neotropics, they have proven to be successful invaders around the world. Some evidence tells us about the most likely first mammal hosts of sand fleas.

As a general rule, adult fleas do not stick closely to a single host species and "accidental" infestations (for example a rabbit flea being found on a predator such as a fox) are not uncommon. You, yourself, may have suffered the bites of cat and dog fleas, since adult fleas can survive on most mammals long enough to get a quick blood meal without being groomed and then jump off. For most fleas, nest or den conditions (where eggs are deposited and larvae develop and pupate) tend to be more important than host species in determining flea survival. For sand fleas in South America, the center of their distribution, usual hosts are armadillos, opossums, and wild rodents. Infections occur on the feet, base of the tails, bellies, and ears of these animals – sites that come into contact with sandy soil as they dig and burrow. These observations have led to an interesting idea – weird sand fleas might not be an outlier in the order but may be the family from which other fleas evolved.

Sand Fleas in Extinct Mammals

Based on family tree analysis, *Tunga* fleas evolved to infect ancient mammals, particularly armadillos, anteaters, and sloths on Gondwanaland (South America, Africa, Australia, Antarctica). These mammals used to be grouped together as "edentates" – mammals with few or no teeth, particularly nipping incisors. Could it be that sand fleas infected these hosts because they could not easily be removed? Fossil bony body plates (nine million years old) from an extinct species of armadillo were found with pea-sized lesions with pores that were likely caused by neosomes of sand fleas. This fossil evidence suggests that endoparasitic sand fleas, with their strange globular bodies inside host skin, are very primitive fleas – even though they also have very unique and specialized lives. Perhaps as mammals developed better grooming methods, incisor teeth, and agile paws with nails, fleas were forced to diversify and to develop the adaptations (flat bodies, combs, spines) that allow them to be ectoparasites in the pelage. At this time, they could then infest a wide variety of den-dwelling mammals, especially the mammalian hosts that had also diversified the most – rodents.

It is incredible to imagine that elephant-sized (1000 kg or 2200 lbs) ground sloths (*Megatherium*) and small car-sized Glyptodonts (extinct armadillos), which excavated tunnels 4 m (13') wide, 2 m (6.6') tall, and 100 m (328') long in the sandy grasslands of South America millions of years ago, were suffering from infections by specialized but evolutionarily primitive *Tunga*-like fleas, which burrowed into their skins (Fig. 23). The tunnels and dens of these extinct mammals probably offered ideal environmental conditions for the fleas to complete their life cycles. Today, living sloths are smaller than their extinct relatives, are very arboreal, and seldom come into contact with the ground (apart from when they defecate, as you may recall from chapter 13). No painful sand flea infections of modern sloths have been reported.

Today, fleas and flea-borne infections are on the rise, and the geographic ranges of fleas are changing and expanding. Evolutionary history shows us how adaptable and resilient most fleas are, even their weird cousin the sand flea.

Figure 23. *Glyptodon* – the large, extinct armadillo – an early host of fleas. Source: Photo © NPL–DeA Picture Library / Bridgeman Images.

Counterintuitively, it is possible that what many people think of as the epitome of ectoparasites, fleas, evolved to live inside giant burrowing animals but went on to very successfully colonize the fur of many mammals and even the nests of birds. When tourists get infected with tungiasis while on a Caribbean holiday, they curse these weird and painful parasites. Perhaps their agony will be eased a little to know that they are hosting parasites that are long-time heirloom associates of ours, dating back to our earliest evolution as mammals. But somehow, I doubt it.

Conclusion: The Greatest Show on Earth

In 2009, British biologist Richard Dawkins wrote a wonderful book that described the evidence for the evolutionary history of life, *The Greatest Show on Earth*. Based on Dawkins's view of life, I would argue that parasites should take their place in the center ring of this show – call it a flea circus if you like.

Parasitism has a prominent role in the history of life. For example, in all cells except bacteria, there reside mitochondria that are responsible for cellular respiration – without mitochondria, we would not be able to use oxygen. It has been theorized that these organelles were originally free-living aerobic (able to use oxygen to create the chemical energy that drives our cells) bacteria, which invaded cells to become endosymbionts (meaning that our cells cannot live without them, and mitochondria cannot reproduce without our cells). No one knows how this relationship started, but I like to think that the ancient bacteria that today are mitochondria originally invaded host cells as parasites. If true, our very bodies (and the bodies of all organisms except bacteria) include artifacts that used to be parasites. However, even if this ancient relationship was not a parasitic one, there is no doubt that parasites have in the past, and are now, playing major roles in the history of life.

Notions about Parasites

There are many misconceptions about parasites, some even held by professional biologists. Parasites are not rare or uncommon in nature; in fact, parasitism is the most common form of energy acquisition in the natural world – more common than predation, scavenging, or even photosynthesis. Parasite examples occur in all kingdoms of living things (bacteria, archaea, plants, fungi, and animals), and even some mammals (such as vampire bats) can be considered parasites of other mammals. Every species of mammal has its own particular suite of parasites living in it and on it, and although most individuals each harbor only a few species of parasites, some hosts are infected and infested by populations of thousands. Undoubtedly, parasites are an integral part of every mammal's biology.

Often, biology is a fuzzy, less precise science than physics or chemistry, and parasitism cannot be absolutely defined – it is a moving target due to evolution, ecology, and chance. Opposite to what you may think, sometimes the association between a parasite and its host provides benefits. For instance, brain worms in white-tailed deer may act as a weapon of competition for the host in its struggle against competitors, allowing deer to invade and thrive in new habitats. Tapeworm and roundworm infections can be better than immune-suppressant drugs such as corticosteroids at reducing colitis and inflammatory bowel diseases.

Another misconception is that natural selection will always lead to a benign compromise between host and parasites, one in which the partners adapt in order to tolerate each other. This idea suggests that, the longer hosts and parasites have lived together, the more harmless the interaction will be. But as we have seen, the roundworm *Trichinella*'s transmission between hosts is facilitated by causing debilitation and death to hosts – in mammals, it can be a horrible, painful killer. As a result, some scavengers will avoid using carnivore carcasses because of the danger of acquiring killer parasites.

Even within one group of animals, like moths, the nature of the parasite-host relationship can vary greatly. Some moths are parasites, piercing blood vessels and drinking host blood and irritating lacrimal ducts in order to drink tears, while others, like the specialized moths living on sloths, are crucially important to supply nourishment for their hosts. This complex symbiotic system ties together sloths, moths, fungi, photosynthetic algae, predatory birds, mammals, and even tropical rainforest trees into one tangled web of interactions. If we can reach any general conclusion, therefore, it is that parasites are wonderfully unpredictable.

Humans Are Mammals Too

Although this book is specifically about wild mammals and their parasites, these lessons also translate to human infections, as we share many parasites with wild and domestic animals. Some parasites act as heirlooms, passed from generation to generation, and tell us a great deal about host evolution, our closest relatives, and when we diverged; others are souvenirs acquired during biogeographical travels or parasites acquired from distantly related hosts that happened to live in the same habitat. Pinworms are roundworm parasites that reveal remarkable evidence of mammals rafting across oceans – not just once but several times – and show how parasites can spread between unrelated hosts like primates and squirrels that live in similar habitats. Head, body, and pubic lice tell us stories about our evolution, our relatives, and our coexistence with other hominids.

Zoonoses (infections of animals that spill over into humans) now account for about 60 percent of emerging diseases, and most of these (72 percent) come from wildlife. It is possible that the coronavirus responsible for our devastating COVID-19 pandemic originated in wild mammals, perhaps bats. The relentless human population expansion results in encroachment into wildlife habitat, and shrinking natural areas leads to increased use of urban areas by wildlife like raccoons, deer, and coyotes. Urban wildlife have parasites that cause dangerous, even deadly human diseases such as visceral larva migrans and hydatid disease. As humans encroach further into previously wild spaces, and as wild animals are driven increasingly into urban areas, we are going to have to develop strategies for dealing with more common outbreaks of zoonotic infections.

Human alteration of habitats, such as converting forests to grasslands, can change the abundance and distribution of intermediate hosts like rodents and entrench zoonotic infections such as hydatid disease or Lyme disease. The relationship between mammal parasites and humans, however, is not just one way. For example, monkeys in South America are infested by lice that they got from us, and coyotes die horrible deaths from sarcoptic mange that we intentionally gave to them.

Mammal parasites have evolved in concert with their hosts, sometimes in an "arms race," where parasites must counteract a battery of host defenses. Today, blood flukes have the upper hand in their competition with their hosts. They have not only evolved ways to survive in the eye of a fierce host immunological storm but have taken over this defense and used it to their own advantage. Hopefully, as we learn more about these parasites, they will give us ideas about methods to control autoimmune diseases like asthma, Crohn's disease, and Addison's disease. Vampire bats have not only coevolved with their mammalian hosts and adapted to the unique bloody diet they follow but have also developed an obligate relationship with a distinctive community of microbes that live in their guts, which potentially could lead us to learn how to use the microbiome for medical treatments.

Parasites: Unseen Agents That Control Nature

Parasites reveal much about host ecology and illustrate patterns that apply to life generally. For example, larger hosts tend to be infected by larger parasites – the largest predator on the planet has the largest species of tapeworm and roundworm that we are aware of. If only we had the opportunity to examine the massive dinosaurs that once roamed our planet, who knows what we would discover. Perhaps parasites and diseases played some role in their extinction, along with an asteroid, and provided the ecological opportunity that mammals took advantage of to diversify.

Parasites impact mammalian behaviors. Antlers, horns, tusks, and other sexual ornaments that characterize many mammals are used for displays but can also confer valuable information to potential mates about disease-resistant genes and parasite infections. It is even possible that some fur colors and patterns, like the stripes on zebras, are due to parasites. What other interesting behaviors, including those of our primate relatives, have something to do with parasites and diseases? The parasitic, blood-feeding lifestyle of vampire bats has led to their social behavior and to sharing food and allogrooming with relatives and even with non-relatives – a type of reciprocal altruism. Could certain human behaviors and characters such as our nakedness and some social behaviors be due to parasites?

Behaviors involving predation may also be affected by parasites. Parasites can change the normal behaviors of intermediate hosts, in many cases causing them to commit suicide by being eaten by mammalian predators – or at least making them easier targets for predators. For instance, muskrats, beavers, wild pigs, and rats can gain more food calories by eating prey that are infected with immature parasites without paying the cost of losing much energy due to infections with adult acanthocephalan parasites, which live in their digestive systems. These kinds of alterations to normal host behaviors affect the dynamics of trophic interactions that shape entire ecosystems.

Even without being involved in predator-prey relationships, parasites can interfere with normal mammalian behaviors that affect ecosystems. For instance, flatworms living in the brains of whales can disrupt social communications, leading to mass stranding events that kill hundreds of animals and affect marine ecosystems by reducing the density of predators and increasing biomass for scavengers.

Curiously, the mechanism of sexual reproduction itself may have been driven by parasites. Sexual reproduction requires organisms to undergo the complex process of meiotic cell division, a process that takes time and often is associated with genetic accidents – for instance, unequal distributions of chromosomes and parts of chromosomes. Biologists suggest that the selective pressure responsible for the evolution of meiosis and sex is the need to produce genetically variable offspring that can cope with changing environmental challenges. Studies of snails that can reproduce sexually or asexually showed that sexual reproduction occurred when snails were exposed to flatworm parasites similar to blood flukes.[1] If the main reason for sex is to produce genetically variable offspring, then parasites could be a greater selection force for this goal even than physical environmental factors (like droughts or cold) because parasites are so common, abundant, and can generate genetically variable strains as they reproduce. Perhaps parasites were the main driving factor – the main environmental challenge – that led organisms to evolve lengthy and dangerous meiotic cell division and, thus, sexual reproduction.

If the meaning of life is reproduction, then parasites live life to its fullest. Most parasites reproduce asexually and can generate thousands to millions of infective stages. However, parasite sexual reproduction can also be important. Mammalian blood flukes are unusual flatworms because they have separate sexes, and the two sexes are very different in size and shape, seemingly driven by the unique challenges of the habitat in the host where they occur. This lifestyle demands a division of labor between the two sexes. The sexual differentiation by blood flukes has led to genetically variable offspring that can cope with ever-altering host defenses. In fact, these worms have become so adept at avoiding host immune responses that they have even managed to take advantage of these defenses in order to sexually mature, produce eggs, get their offspring out of their host, and prevent hosts from getting "super infected," thus allowing the infected mammals to live with their parasites for years. At the other end of the spectrum, many parasites like giant liver flukes and hydatid-forming tapeworms reproduce asexually to massively increase their numbers, flooding the environment and ensuring that some offspring survive the dangerous perils necessary for transmission to new hosts.

I think that, if Charles Darwin were alive today, he would be thrilled with the knowledge we have learned about parasites – there are no better illustrations of the power of natural selection and the creative force of evolution in the biological world. Powerful natural selection has forced parasites to evolve suites of specializations for finding hosts, getting into or onto them, holding on once they are there, reproducing, getting eggs or larvae out, and finding ways to get into the next generation of hosts. Many of these elaborate adaptations, such as taking over the genetic control of cells or altering infected host behaviors, are so incredible that they seem to belong more to the world of science fiction than to the science of biology.

Our Changing World

Like all other ecological interactions and organisms today, parasites respond to environmental changes and challenges. Some, like brain worms, liver flukes, fleas, and ticks, may increase transmission and expand their ranges. Others, like *Trichinella nativa*, may become restricted, replaced by other species, or go extinct. Because all mammals are infected with parasites (usually by more than one species) and because coevolution has led to specialization and cospeciation, every time a host goes extinct, one or more parasites likely go extinct as well.

Sadly, we are currently experiencing a mass extinction event that rivals those of the past, like the one at the end of the Cretaceous, which resulted in the end of dinosaurs. Today's mass extinction, however, is not caused by some uncontrollable factor like an asteroid striking the planet. It is being caused by us. Our wanton destruction of natural biomes such as tropical rainforests through unsustainable extraction of resources, our urbanization of wild spaces, our unmitigated pollution of oceans with plastics and other garbage, and our failure to restrain our production of greenhouse gases, which are changing the climate, are all decimating the biodiversity of Earth. This human-induced mass extinction event is probably causing more parasite extinctions than host extinctions. But before we think that parasite extinction is good news, reflect that parasites are integral strands in ecological webs. Loss of parasites may result in unpredictable consequences and environmental imbalances that have negative impacts on entire ecosystems, which can directly impact human well-being when novel diseases and infections like COVID-19 appear in ecosystems and spread into humans. An entirely likely outcome of the expansion of distributions of some animals, like urban coyotes and racoons, will be significant increases in cases of hydatid disease and ascarid roundworm infections in humans. Unfortunately, this experiment is one for which we will only get the results many years from now.

Naturally, parasites affect habitat use, abundance, and distributions of hosts. Raccoon latrines that vary in location from the bases of boulders to the tops of hills are avoided by many mammals and birds that recognize these sites are death traps due to roundworms. Grazing mammals avoid areas where fecal contamination could mean parasite infections, and bats will give up important roosting sites when populations of their hyperparasitic, blood-feeding bat bugs get too high. Parasites can also affect geographic ranges and land use by wild mammals.

How global climate change will impact parasites and their hosts is a big black box. As the arctic climate becomes warmer, mammals like grizzly bears may expand their range northward and bring new parasites with them, and marine mammals such as walruses will be forced to change diets and ranges, causing more cases of human trichinosis. As agriculture in higher latitudes becomes more feasible, domesticated animals may bring new infections, which could be devastating to immunologically naive native wildlife like muskoxen and barren ground caribou. Droughts may concentrate wildlife around predictable sources of water, which will facilitate parasite transmission and act as disease hot spots, and a warmer climate will increase populations of fleas and ticks, directly resulting in host population die-offs (for example, in moose with winter ticks) and more bacterial and viral diseases in humans like Lyme disease and yellow fever.

Have I Exaggerated?

As a parasitologist, I am often accused of exaggerating the role that parasites play in the grand scheme of nature. Surely, it is argued, parasites are too rare or infrequent, or too insignificant with regard to the effects they have on hosts, to be as important as I claim. Other, more obvious factors, like predators or competitors or weather events and climate, must certainly have been stronger selection pressures that have shaped the anatomy, behavior, and ecology of hosts such as wild mammals. Yet when biologists look deeply into the lives of organisms, they often discover there is an unseen world of mutualists and parasites that are working behind the scenes, constantly intruding into the lives of their hosts – and because parasitism is so successful, it is no wonder that so many organisms have taken up a parasitic existence as a way to get the resources they need to reproduce.

I believe that this book has introduced you to a world that most people are completely unaware of. Animals we usually think of as parasites – fleas, flukes, roundworms, tapeworms, ticks, mites, and lice – shape the lives and evolution of wild mammals, as do organisms we usually don't think of as parasites, such as beetles, moths, and flies. Biologists who study parasitic organisms are lucky – they get front row seats to a vital part of the greatest show on Earth. The more we learn about parasites, the more we learn about mammals, about ecosystems, about evolution and the history of life, about the beautiful, complex tapestry of nature, and even about ourselves.

Notes

1. Pinworms, Primates, and Porcupines: How Parasites Traveled the World

1　Lashaki et al. 2023.
2　Dudlová et al. 2018.

2. Stone Cold Killers: *Trichinella* in the Arctic

1　For example, Campbell 1988.
2　Springer et al. 2017.
3　Lowry and Fay 1984.
4　Jex 2016.

3. Who's Your Daddy? Lice on Great Apes

1　Batterberry and Batterberry 1977, p. 81.

5. Beetles and Beavers

1　Quoted in Gould 1993.
2　Goodman et al. 2017.

6. Stranded Whales: A Fluke Accident?

1　Becker 2017.
2　Rinehart et al. 1999.

3 See Degollada et al. 2002.
4 See, for example, Bashey 2015.

7. How the Zebra Got Its Stripes

1 Darwin 1871.
2 Wallace 1889.
3 For example, Caro et al. 2014; Kingdon 1984.
4 Caro 2016; Larison et al. 2015.
5 Horváth et al. 2018.
6 Auty et al. 2016.
7 Larison et al. 2015.
8 Caro 2016.
9 Takács et al. 2022.
10 Kojima et al. 2019.

8. Ornaments and Parasites

1 See, for example, Folstad et al. 1996.
2 Folstad et al. 1996.
3 Luzón et al. 2008.
4 Luzón et al. 2008.
5 Mulvey and Aho 1993.
6 Ezenwa et al. 2012.

10. Your Brain on Worms: Nature's Biological Weapon

1 Prestwood et al. 1974.
2 Kutz et al. 2012.
3 Kutz et al. 2001.

11. The Tale of the Tape: The World's Longest Parasite

1 See, for example, Morand and Poulin 2002.
2 Tsai et al. 2003.

12. Death by Raccoon

1 Grassi 1888.
2 Rainwater et al. 2017.
3 Fox et al. 1985.
4 Weinstein et al. 2017.

13. Moths, Sloths, Tears, and Blood

1 Waage 1979.

14. The Manchurian Parasite

1. Holmes and Bethel 1972.
2. Bethel and Holmes 1973.
3. Moore 1983.
4. Moore 2002.
5. Moore and Freehling 2002.
6. Moore 2002.
7. Berdoy et al. 1995; Servick 2020; Webster 2007.
8. See, for example, Vyas 2015.
9. See, for example, Servick 2020.

15. A Ghost of a Chance

1. Samuel 2004.

16. Sex and the Single Schistosome

1. World Health Organization 2017.
2. World Health Organization 2023.
3. Zhang et al. 2001.
4. Kuperschmidt 2018.

17. The Trickster: Coyotes and Their Parasites

1. Grinder and Krausman 2001.
2. Liccioli et al. 2012.
3. Luong et al. 2020.
4. Gesy et al. 2014; Liccioli et al. 2012; Luong et al. 2020.
5. Holmes and Podesta 1968.
6. Samuel et al. 1978.
7. Luong et al. 2020.
8. Holmes et al. 1971.
9. Houston et al. 2021.
10. Luong et al. 2020.
11. City of Edmonton, n.d.

18. Fleas: The Inside Story

1. Whiting et al. 2008.
2. Stewart 2011.
3. Linardi and Moreira 2014.

Conclusion: The Greatest Show on Earth

1. Jokela et al. 2009; Lively 1987.

References and Additional Readings

Introduction: Wolves and Worms

Bryan, R.T., and P.M. Shantz. 1989. Echinococcosis (hydatid disease). J. Am. Vet. Med. Assoc. 195:1214–17.

Joly, D.O., and F. Messier. 2004. The distribution of *Echinococcus granulosus* in moose: Evidence for parasite-induced vulnerability to predation by wolves. Oecologia 140:586–90. https://doi.org/10.1007/s00442-004-1633-0.

Lymbery, A.J., and R.C.A. Thompson. 1996. Species of *Echinococcus*: Pattern and process. Parasitol. Today 12:486–91. https://doi.org/10.1016/s0169-4758(96)10071-5.

McManus, D.P., and J.D. Smyth. 1986. Hydatidosis: Changing concepts in epidemiology and speciation. Parasitol. Today 2:163–8. https://doi.org/10.1016/0169-4758(86)90147-X.

McTaggart-Cowan, I. 1951. The diseases and parasites of big game mammals of western Canada. Proc. 5th Ann. Game Conv., pp. 37–64.

Rausch, R.L. 1952. Hydatid disease in boreal regions. Arctic 5:157–74.

Smith, D.W., R.O. Peterson, and D.B. Houston. 2003. Yellowstone after wolves. BioScience 53:330–40. https://doi.org/10.1641/0006-3568(2003)053[0330:YAW]2.0.CO;2.

Smyth, J.D. 1964. The biology of the hydatid organisms. Adv. Parasitol. 2:169–219. https://doi.org/10.1016/S0065-308X(08)60588-6.

Thomson, R.C.A., and A.J. Lymbery (eds.). 1986. The biology of *Echinococcus* and hydatid disease. London: Allen & Unwin.

Zimmer, C. 2000. Parasite rex: Inside the bizarre world of nature's most dangerous creatures. New York: Simon and Schuster.

1. Pinworms, Primates, and Porcupines: How Parasites Traveled the World

Brooks, D.R., and D.R. Glen. 1982. Pinworms and primates: A case study in coevolution. Proc. Helminthol. Soc. Wash. 49:76–85.

Cameron, T.W. 1929. The species *Enterobius* Leach, in primates. J. Helminthol. 7:161–82.

Dudlová, A., P. Juriš, P. Jarčuška, Z. Vasilková, V. Varcová, M. Sumková, and V. Krčméry. 2018. The incidence of pinworm (*Enterobius vermicularis*) in pre-school and school aged children in eastern Slovakia. Helminthologia 55:275–80. https://doi.org/10.2478/helm-2018-0030.

Foitová, I., K. Civáňová, V. Baruš, and W. Nurcahyo. 2014. Phylogenetic relationships between pinworms (Nematoda: Enterobiinae) parasitising the critically endangered orangutan, according to the characterization of molecular genomic and mitochondrial markers. Parasitol. Res. 113:2455–66. https://doi.org/10.1007/s00436-014-3892-y.

Hugot, J.-P. 1999. Primates and their pinworm parasites: The Cameron hypothesis revisited. Syst. Biol. 48:523–46. https://doi.org/10.1080/106351599260120.

Hugot, J-P. 2002. New evidence of Hystricognath rodents monophyly from the phylogeny of their pinworms. In: R.D.M. Page (ed.), Tangled trees: Phylogeny, cospeciation, and coevolution, pp. 144–73. Chicago: Univ. Chicago Press.

Lashaki, E.K., A. Mizani, S.A. Hosseini, B. Habibi, K. Takerkhani, A. Javadi, A. Taremiha, and S. Dodangeh. 2023. Global prevalence of enterobiasis in young children over the past 20 years: A systematic review and meta-analysis. Osong Public Health Res. Perspect. 14:441–50. https://doi.org/10.24171/j.phrp.2023.0204.

Nakano, T., M. Okamoto, Y. Ikeda, and H. Hasegawa. 2006. Mitochondrial cytochrome c oxidase subunit 1 gene and nuclear rDNA regions of *Enterobius vermicularis* parasitic in captive chimpanzees with special reference to its relationship with pinworms in humans. Parasitol. Res. 100:51–7. https://doi.org/10.1007/s00436-006-0238-4.

2. Stone Cold Killers: *Trichinella* in the Arctic

Asbakk, K., J. Aars, A.E. Derocher, O. Wiig, A. Oksanen, E.W. Born, R. Dietz, C. Sonne, J. Godfroid, and C.M.O. Kapel. 2010. Serosurvey for *Trichinella* in polar bears (*Ursus maritimus*) from Svalbard and the Barents Sea. Vet. Parasitol. 172:256–63. https://doi.org/10.1016/j.vetpar.2010.05.018.

Campbell, W.C. 1988. Trichinosis revisited – Another look at modes of transmission. Parasitol. Today 4:83–6. https://doi.org/10.1016/0169-4758(88)90203-7.

Connell, F.H. 1949. Trichinosis in the Arctic: A review. Arctic 2:98–107. https://www.jstor.org/stable/40506354.

Conover, M.R., and R.M. Vail. 2015. Human diseases from wildlife. Boca Raton (FL): CRC Press.

Deksne, G., Z. Seglina, I. Jahundoviča, Z. Esīte, E. Bakasejevs, G. Bagrade, D. Keidāne, M. Interisano, G. Marucci, D. Tonanzi, E. Pozio, and M. Kirjušina. 2016. High prevalence of *Trichinella* spp. in sylvatic carnivore mammals of Latvia. Vet. Parasitol. 231:118–23. https://doi.org/10.1016/j.vetpar.2016.04.012.

Gilabert, A., and J.D. Wasmuth. 2013. Unravelling parasitic nematode natural history using population genetics. Trends Parasitol. 29:438–48. https://doi.org/10.1016/j.pt.2013.07.006.

Goździk, K., I.M. Odoevskaya, S.O. Movsesyan, and W. Cabaj. 2017. Molecular identification of *Trichinella* isolates from wildlife animals of the Russian Arctic territories. Helminthologia 54:11–16. https://doi.org/10.1515/helm-2017-0002.

Jex, C. 2016. Grizzly-polar bear hybrids spotted in Canadian Arctic. ScienceNordic, June 6, 2016. https://www.sciencenordic.com/denmark-evolution-greenland-science-special/grizzly-polar-bear-hybrids-spotted-in-canadian-arctic/1434185.

Lowry, L.F., and F.H. Fay. 1984. Seal eating by walruses in the Bering and Chukchi Seas. Polar Biology 3:11–18. https://doi.org/10.1007/BF00265562.

Polley, L., E. Hoberg, and S. Kutz. 2010. Climate change, parasites and shifting boundaries. Acta Vet. Scand. 52(Suppl. 1):S1. https://doi.org/10.1186/1751-0147-52-S1-S1.

Rausch, R.L., J.C. George, and H.K. Brower. 2007. Effect of climate warming on the Pacific walrus, and potential modification of its helminth fauna. J. Parasitol. 93:1247–51. https://doi.org/10.1645/GE-3583CC.1.

Springer, Y.P., S. Casillas, K. Helfrich, D. Mocan, M. Smith, G. Arriaga, L. Mixson, L. Castrodale, and J. McLaughlin. 2017. Two outbreaks of trichinellosis linked to consumption of walrus meat – Alaska, 2016 –2017. MMWR Morb. Mortal. Wkly. Rep. 66:692–6. https://doi.org/10.15585/mmwr.mm6626a3.

Stefansson, V. 1948. Trichinosis from polar bear meat. Arctic 1:144.

3. Who's Your Daddy? Lice on Great Apes

Ashfaq, M., S. Prosser, S. Nasir, M. Masood, S. Ratnasingham, and P.D.N. Hebert. 2015. High diversity and rapid diversification in the head louse, *Pediculus humanus* (Pediculidae: Phthiraptera). Sci. Rep. 5:14188. https://doi.org/10.1038/srep14188.

Batterberry, M., and A. Batterberry. 1977. Mirror, mirror: A social history of fashion. New York: Holt, Rinehart, and Winston.

Boutellis, A., L. Abi-Rached, and D. Raoult. 2014. The origin and distribution of human lice in the world. Infect. Genet. Evol. 23:209–17. https://doi.org/10.1016/j.meegid.2014.01.017.

Dholakia, S., J. Buckler, J.P. Jeans, A. Pillai, N. Eagles, and S. Dholakia. 2014. Pubic lice: An endangered species? Sex Transm. Dis. 41:388–91. https://doi.org/10.1097/olq.0000000000000142.

Ditrich, H. 2017. The transmission of the Black Death to western Europe: A critical review of the existing evidence. Mediterr. Hist. Rev. 32:25–39. https://doi.org/10.1080/09518967.2017.1314920.

Drali, R., L. Abi-Rached, A. Boutellis, F. Djossou, S.C. Barker, and D. Raoult. 2016. Host switching of human lice to new world monkeys in South America. Infect. Genet. Evol. 39:225–31. https://doi.org/10.1016/j.meegid.2016.02.008.

Durant, W. 1950. The age of faith (The story of civilization, vol. 4). New York: Simon and Schuster.

Light, J.E., and D.L. Reed. 2009. Multigene analysis of phylogenetic relationships and divergence times of primate sucking lice (Phthiraptera: Anoplura). Mol. Phylogenet. Evol. 50:376–90. https://doi.org/10.1016/j.ympev.2008.10.023.

Pagel, M., and W. Bodmer. 2003. A naked ape would have fewer parasites. Proc. Roy. Soc. B. Biol. Sci. 270(Supp 1):S117–19. https://doi.org/10.1098/rsbl.2003.0041.

Rantala, M.J. 1999. Human nakedness: Adaptation against ectoparasites? Int. J. Parasitol. 29:1987–9. https://doi.org/10.1016/s0020-7519(99)00133-2.

Reed, D.L., J.E. Light, J.M. Allen, and J.J. Kirchman. 2007. Pair of lice lost or parasites regained: The evolutionary history of anthropoid primate lice. BMC Biol. 5:7. https://doi.org/10.1186/1741-7007-5-7.

Smith, V.S. 2004. Lousy phylogenies: Phthiraptera systematics and the antiquity of lice. Proc. 1st Dresden Meeting on Insect Phylogeny, Entomol. Abhandlungen 61:150–1.

Veracx, A., and D. Raoult. 2012. Biology and genetics of human head and body lice. Trends Parasitol. 28:563–71. https://doi.org/10.1016/j.pt.2012.09.003.

Weiss, R.A. 2007. Lessons from naked apes and their infections. J. Med. Primatol. 36:172–9. https://doi.org/10.1111/j.1600-0684.2007.00235.x.

Weiss, R.A. 2009. Apes, lice and prehistory. J. Biol. 8:20. https://doi.org/10.1186/jbiol114.

4. Giants Crawl among Us: Giant Liver Flukes

Králová-Hromadová, I., L. Juhásová, and E. Bazsalovicsová. 2016. The giant liver fluke, *Fascioloides magna*: Past, present and future research. Cham (CH): Springer Int'l. Pub.

Malcicka, M. 2015. Life history and biology of *Fascioloides magna* (Trematoda) and its native and exotic hosts. Ecol. Evol. 5:1381–97. https://doi.org/10.1002/ece3.1414.

Mulvey, M., and J.M. Aho. 1993. Parasitism and mate competition: Liver flukes in white-tailed deer. Oikos 66:187–92. https://doi.org/10.2307/3544804.

Novobilsky, A., E. Horácková, L. Hirtová, D. Modrý, and B. Koudela. 2007. The giant liver fluke *Fascioloides magna* (Bassi 1875) in cervids in the Czech Republic and potential of its spreading to Germany. Parasitol. Res. 100:549–53. https://doi.org/10.1007/s00436-006-0299-4.

Pybus, M.J. 2001. Liver flukes. In: W.M. Samuel, M.J. Pybus, and A.A. Kocan (eds.), Parasitic diseases of wild mammals. 2nd ed., pp. 121–49. Ames (IA): Iowa State Univ. Press.

Roberts, L.S., J. Janovy, Jr., and S. Nadler. 2013. Foundations of parasitology. 9th ed. New York: McGraw Hill.

5. Beetles and Beavers

Brown, J.H. 1981. Two decades of homage to Santa Rosalia: Toward a general theory of biodiversity. Amer. Zool. 21:877–88. https://doi.org/10.1093/icb/21.4.877.

Brunke, A., A. Smetana, D. Carruthers-Lay, and J. Buffam. 2017. Revision of *Hemiquedius* Casey (Staphylinidae, Staphylininae) and a review of beetles dependent on beavers and muskrats in North America. ZooKeys 702:27–43. https://doi.org/10.3897/zookeys.702.19936.

Farrell, B.D. 1998. "Inordinate fondness" explained: Why are there so many beetles? Science 281:555–9. https://doi.org/10.1126/science.281.5376.555.

Goodman, G., A. Meredith, S. Girling, F. Rosell, and R. Campbell-Palmer. 2017. Outcomes of a "One Health" monitoring approach to a five-year beaver (*Castor fiber*) reintroduction trial in Scotland. EcoHealth 14(Supp 1):S139–43. https://doi.org/10.1007/s10393-016-1168-y.

Gould, S.J. 1993. A special fondness for beetles. Nat. Hist. 102:4–8.

Horn, S., W. Durka, R. Wolf, A. Ermala, A. Stubbe, M. Stubbe, and M. Hofreiter. 2011. Mitochondrial genomes reveal slow rates of molecular evolution and the timing of speciation in beavers (*Castor*), one of the largest rodent species. PLoS One 6:e14622. https://doi.org/10.1371/journal.pone.0014622.

Hutchinson, G.E. 1959. Homage to Santa Rosalia or why are there so many kinds of animals? Am. Nat. 93:145–59. https://doi.org/10.1086/282070.

Janzen, D.H. 1963. Observations on populations of adult beaver-beetles, *Platypsyllus castoris* (Platypsyllidae: Coleoptera). Pan-Pac. Entomol. 39:215–28.

Jorgensen, D. 2015. Conservation implications of parasite co-reintroduction. Conserv. Biol. 29:602–4. https://doi.org/10.1111/cobi.12421.

Peck, S.B. 2006. Distribution and biology of the ectoparasitic beaver beetle *Platypsyllus castoris* Ritsema in North America (Coleoptera: Leiodidae: Platypsyllinae). Insecta Mundi 20:87–96.

Smith, D.M., and J.D. Marcot. 2015. The fossil record and macroevolutionary history of beetles. Proc. Roy. Soc. Lond. B. Biol. Sci. 282:20150060. https://doi.org/10.1098/rspb.2015.0060.

Théodoridès, J. 1950. The parasitological, medical and veterinary importance of Coleoptera. Acta Tropica 7:48–60.

University of California – Los Angeles. 2012. Why are there so many species of beetles and so few crocodiles? ScienceDaily, August 28. https://www.sciencedaily.com/releases/2012/08/120828171744.htm.

Whitaker, J.O., Jr., A. Fain, and G.S. Jones. 1989. Ectoparasites from beavers from Massachusetts and Maine. Intl. J. Acarol. 15:153–62. https://doi.org/10.1080/01647958908683841.

6. Stranded Whales: A Fluke Accident?

Bashey, F. 2015. Within-host competitive interactions as a mechanism for the maintenance of parasite diversity. Philos. Trans. R. Soc. Lond. B. Biol. Sci. 370:20140301. https://doi.org/10.1098/rstb.2014.0301.

Becker, R. 2017. New Zealand just experienced its largest whale stranding in decades. The Verge, February 11, 2017. https://theverge.com/2017/2/11/14587770.

Costidis, A., and S.A. Rommel. 2012. Vascularization of air sinuses and fat bodies in the head of the bottlenose dolphin (*Tursiops truncatus*): Morphological implications on physiology. Front. Physiol. 3:243. https://doi.org/10.3389/fphys.2012.00243.

Dailey, M.D. 2005. Parasites of marine mammals. In: K. Rohde (ed.), Marine parasitology, pp. 408–14. Clayton South Vic (AUS): CSIRO.

Dailey, M.D., and W.A. Walker. 1978. Parasitism as a factor (?) in single strandings of southern California cetaceans. J. Parasitol. 64:593–6. https://doi.org/10.2307/3279939.

Degollada, E., M. André, M. Arbelo, and A. Fernández. 2002. Incidence, pathology and involvement of *Nasitrema* species in odontocete strandings in the Canary Islands. Vet. Rec. 150:81–2. https://doi.org/10.1136/vr.150.3.81.

Ebert, M.B., and A.L.S. Valentere. 2013. New records of *Nasitrema atenuatta* and *Nasitrema globicephalae* (Trematoda: Brachycladiidae) Neiland, Rice and Holden, 1970 in delphinids from South Atlantic. Check List 9:1538–40. https://doi.org/10.15560/9.6.1538.

Fernández, A., J.F. Edwards, F. Rodrígez, A. Espinosa de los Monteros, P. Herráez, P. Castro, J.R. Jaber, V. Martin, and M. Arbelo. 2005. "Gas and fat embolic syndrome" involving a mass stranding of beaked whales (family Ziphiidae) exposed to anthropogenic sonar signals. Vet. Path. 42:446–57. https://doi.org/10.1354/vp.42-4-446.

Hermosilla, C., L.M.R. Silva, R. Prieto, S. Kleinertz, A. Taubert, and M. Silva. 2015. Endo- and ectoparasites of large whales (Cetartiodactyla: Balaenopteridae, Physeteridae): Overcoming difficulties in obtaining appropriate samples by non- and minimally-invasive methods. Intl. J. Parasitol. Parasites Wildl. 4:414–20. https://doi.org/10.1016%2Fj.ijppaw.2015.11.002.

Neiland, K.A., D.W. Rice, and B.L. Holden. 1970. Helminths of marine mammals, I. The genus *Nasitrema*, air sinus flukes of delphinid cetaceans. J. Parasitol. 56:305–16. https://doi.org/10.2307/3277662.

Oliveira, J.B., J.A. Morales, R.C. González-Barrientos, J. Hernández-Gamboa, and G. Hernández-Mora. 2011. Parasites of cetaceans stranded on the Pacific coast of Costa Rica. Vet. Parasitol. 182:319–28. https://doi.org/10.1016/j.vetpar.2011.05.014.

Phillips, A.C.N., and R. Suepaul. 2017. *Nasitrema* species: A frequent culprit in melon-headed whale (*Peponocephala electra*) strandings in Trinidad. Aquat. Mamm. 43:547–57. https://doi.org/10.1578/AM.43.5.2017.547.

Rhinehart, H.L., C.A. Manire, F.W. Klutzow, and G.D. Bossart. 1999. Brain lesions and clinical signs in two pygmy killer whales (*Feresa attenuata*) associated with *Nasitrema* sp. IAAAM Archive 1999. https://www.vin.com/apputil/content/defaultadv1.aspx?id=3864426.

Weilgart, L.S., and H. Whitehead. 1990. Vocalizations of the North Atlantic pilot whale (*Globicephala melas*) as related to behavioral contexts. Behav. Ecol. Sociobiol. 26:399–402. https://doi.org/10.1007/BF00170896.

7. How the Zebra Got Its Stripes

Auty, H., S. Cleveland, I. Malele, J. Masoy, T. Lembo, P. Bassell, S. Torr, K. Picozzi, and S.C. Welburn. 2016. Quantifying heterogeneity in host-vector contact: Tsetse (*Glossina swynnertoni* and *G. pallidipes*) host choice in Serengeti National Park, Tanzania. PLoS One 11:e0161291. https://doi.org/10.1371/journal.pone.0161291.

Caro, T. 2016. Zebra stripes. Chicago: Univ. Chicago Press.

Caro, T., A. Izzo, R.C. Reiner Jr., H. Walker, and T. Stankowich. 2014. The function of zebra stripes. Nat. Commun. 5:3535. https://doi.org/10.1038/ncomms4535.

Darwin, C.E. 1871. The descent of man and selection in relation to sex. London: John Murray.

Egri, A., M. Blahó, G. Kriska, R. Farkas, M. Gyurkovszky, S. Åkesson, and G. Horváth. 2012. Polarotactic tabanids find striped patterns with brightness and/or polarization modulation least attractive: An advantage of zebra stripes. J. Exp. Biol. 215:736–45. https://doi.org/10.1242/jeb.065540.

Gosling, M. 2017. Biting flies, lions, and the evolution of zebra stripes. Book review. Ecology 98:2227–9. https://doi.org/10.1002/ecy.1866.

Horváth, G., A. Pereszlényi, D. Száz, A. Barta, I. Jánosi, B. Gerics, and S. Åkesson. 2018. Experimental evidence that stripes do not cool zebras. Sci. Rep. 8:9351. https://doi.org/10.1038/s41598-018-27637-1.

How, M.J., and J.M. Zanker. 2014. Motion camouflage induced by zebra stripes. Zoology 117:163–70. https://doi.org/10.1016/j.zool.2013.10.004.

Kingdon, J. 1984. The zebra's stripes: An aid to group cohesion? In: D.W. MacDonald (ed.), Encyclopedia of mammals, pp. 486–7. Crows Nest, AU: Allen and Unwin Press.

Kojima, T., K. Oishi, Y. Matsubara, Y. Uchiyama, Y. Fukushima, N. Aoki, S. Sato, T. Masuda, J. Ueda, H. Hirooka, and K. Kino. 2019. Cows painted with zebra-like striping can avoid biting fly attack. PLoS One. 14:e0223447. https://doi.org/10.1371/journal.pone.0223447.

Larison, B., R.J. Harrigan, H.A. Thomassen, D.I Rubenstein, A.M. Chan-Golston, E. Li, and T.B. Smith. 2015. How the zebra got its stripes: A problem with too many solutions. R. Soc. Open Sci. 2:140452. https://doi.org/10.1098/rsos.140452.

Melin, A.D., D.W. Kline, C. Hiramatsu, and T. Caro. 2016. Zebra stripes through the eyes of their predators, zebras, and humans. PLoS One 11:e0145679. https://doi.org/10.1371/journal.pone.0145679.

Takács, P., D. Száz, M. Vincze, J. Slíz-Balogh, and G. Horváth. 2022. Sunlit zebra stripes may confuse the thermal perception of blood vessels causing the visual unattractiveness of zebras to horseflies. Sci. Rep. 12:10871. https://doi.org/10.1038/s41598-022-14619-7.

Waage, J.K. 1981. How the zebra got its stripes – Biting flies as selective agents in the evolution of zebra coloration. J. Ent. Soc. Sth. Afr. 44:351–8. https://hdl.handle.net/10520/AJA00128789_3800.

Wallace, A.R. 1889. Darwinism: An exposition of the theory of natural selection with some of its applications. London: Macmillan.

8. Ornaments and Parasites

Darwin, C.E. 1871. The descent of man and selection in relation to sex. London: John Murray.

Doucet, S.M., and R. Montgomerie. 2003. Structural plumage and parasites in satin bowerbirds: Implications for sexual selection.

J. Avian Biol. 34:237–42. https://doi.org/10.1034/j.1600-048X .2003.03113.x.

Ezenwa, V.O., L.S. Ekernas, and S. Creel. 2012. Unravelling complex associations between testosterone and parasite infection in the wild. Funct. Ecol. 26:123–33. https://doi.org/10.1111/j.1365-2435.2011.01919.x.

Ezenwa, V.O., and A.E. Jolles. 2008. Horns honestly advertise parasite infection in male and female African buffalo. Anim. Behav. 75:2013–21. https://doi.org/10.1016/j.anbehav.2007.12.013.

Folstad, I., P. Arneberg, and A.J. Karter. 1996. Antlers and parasites. Oecologia 105:556–8. https://doi.org/10.1007/bf00330020.

Gould, S.J. 1974. The origin and function of "bizarre" structures: Antler size and skull size in the "Irish elk," *Megaloceros giganteus*. Evolution 28:191–220. https://doi.org/10.1111/j.1558-5646.1974.tb00740.x.

Hamilton, W.D., and M. Zuk. 1982. Heritable true fitness and bright birds: A role for parasites? Science 218:384–7. https://doi.org/10.1126/science.7123238.

Luzón, M., J. Santiago-Moreno, A. Meana, A. Toledano-Díaz, A. Pulido-Pastor, A. Gómez-Brunet, and A. López-Sebastián. 2008. Parasitism and horn quality in male Spanish ibex (*Capra pyrenaica hispanica*) from Andalucía based on coprological analysis and muscle biopsy. Span. J. Agric. Res. 6:353–61. https://doi.org/10.5424/sjar/2008063-328.

Mulvey, M., and J.M. Aho. 1993. Parasitism and mate competition: Liver flukes in white-tailed deer. Oikos 66:187–92. https://doi.org/10.2307/3544804.

Watve, M.G., and R. Sukumar. 1997. Asian elephants with longer tusks have lower parasite loads. Current Science 72: 885–9. https://www.jstor.org/stable/24100035.

Zahavi, A., and A. Zahavi. 1997. The handicap principle: A missing piece of Darwin's puzzle. New York: Oxford Univ. Press.

9. The Night of the Vampire: Parasitic Mammals and Bat Bugs

Carter, G.G., G.S. Wilkinson, and R.A. Page. 2017. Food-sharing vampire bats are more nepotistic under conditions of perceived risk. Behav. Ecol. 28:565–9. https://doi.org/10.1093/beheco/arx006.

Dick, C.W., and B.D. Patterson. 2006. Bat flies: Obligate ectoparasites of bats. In: S. Morand, B.R. Krasnov, and R. Poulin (eds.), Micromammals and macroparasites: From evolutionary ecology to management, pp. 179–94. London: Springer.

Fenton, M.B. 1983. Just bats. Toronto (ON): Univ. Toronto Press.
Kricher, J.C. 2017. The new neotropical companion. Princeton (NJ): Princeton Univ. Press.
Kunz, T.H. 1982. Roosting ecology of bats. In: T.H. Kunz (ed.), Ecology of bats, pp. 1–55. New York: Plenum Publ.
Lord, R.D. 2007. Mammals of South America. Baltimore (MD): Johns Hopkins Univ. Press.
Patterson, B.D., C.W. Dick, and K. Dittmar. 2007. Roosting habits of bats affect their parasitism by bat flies (Diptera: Streblidae). J. Trop. Ecol. 23:177–89. https://doi.org/10.1017/S0266467406003816.
Rechardt, K., and G. Kerth. 2007. Roost selection and roost switching of female Bechstein's bats (*Myotis bechsteinii*) as a strategy of parasite avoidance. Oecologia 154:581–8. https://doi.org/10.1007/s00442-007-0843-7.
Schutt, B. 2008. Dark banquet: Blood and the curious lives of blood-feeding creatures. New York: Harmony Books.
Stuckey, M.J., B.B. Chomel, E.C. de Fleurieu, A. Aguilar-Setién, H.-J. Boulouis, and C.-C. Chang. 2017. *Bartonella*, bats and bugs: A review. Comp. Immunol. Microbiol. Infect. Dis. 55:20–9. https://doi.org/10.1016/j.cimid.2017.09.001.
Zepeda Mendoza, M.L., Z. Xiong, M. Escalera-Zamudio, A.K. Runge, J. Thézé, D. Streicker, H.K. Frank, E. Loza-Rubio, S. Liu, O.A. Ryder, J.A. Samaniego Castruita, A. Katzourakis, G. Pacheco, B. Taboada, U. Löber, O.G. Pybus, Y. Li, E. Rojas-Anaya, K. Bohmann, A.C. Baez, C.F. Arias, S. Liu, A.D. Greenwood, M.F. Bertelsen, N.E. White, M. Bunce, G. Zhang, T. Sicheritz-Pontén, and M.P.T. Gilbert. 2018. Hologenomic adaptations underlying the evolution of sanguivory in the common vampire bat. Nat. Ecol. Evol. 2:659–68. https://doi.org/10.1038/s41559-018-0476-8.

10. Your Brain on Worms: Nature's Biological Weapon

Anderson, R.C. 1963. The incidence, development, and experimental transmission of *Pneumostrongylus tenuis* Doughtery (Metastrongyloidea: Protostrongylidae) of the meninges of the white-tailed deer (*Odocoileus virginianus borealis*) in Ontario. Can. J. Zool. 41:775–92. https://doi.org/10.1139/z63-049.
Asmundsson, I.M., J.A. Mortenson, and E.P. Hoberg. 2008. Muscleworms, *Parelaphostrongylus andersoni* (Nematoda: Protostrongylidae),

discovered in Columbia white-tailed deer from Oregon and Washington: Implications for biogeography and host associations. J. Wildl. Dis. 44:16–27. https://doi.org/10.7589/0090-3558-44.1.16.

Carreno, R.A., L.A. Durdan, D.R. Brooks, A. Abrams, and E.P. Hoberg. 2001. *Parelaphostrongylus tenuis* (Nematoda: Protostrongylidae) and other parasites of white-tailed deer (*Odocoileus virginianus*) in Costa Rica. Comp. Parasitol. 68:177–84.

Feldman, R.E., M.J.L. Peers, R.S.A. Pickles, D. Thornton, and D.L. Murray. 2017. Climate driven range divergence among host species affects range-wide patterns of parasitism. Glob. Ecol. Conserv. 9:1–10. https://doi.org/10.1016/j.gecco.2016.10.001.

Jenkins, E.J., G.D. Appleyard, E.P. Hoberg, B.M. Rosenthal, S.J. Kutz, A.M. Veitch, H.M. Schwantje, B.T. Elkin, and L. Polley. 2005. Geographic distribution of the muscle-dwelling nematode *Parelaphostrongylus odocoilei* in North America, using molecular identification of first-stage larvae. J. Parasitol. 91:574–84. https://doi.org/10.1645/ge-413r.

Jenkins, E.J., A.M. Veitch, S.J. Kutz, E.P. Hoberg, and L. Polley. 2006. Climate change and the epidemiology of protostrongylid nematodes in northern ecosystems: *Parelaphostrongylus odocoilei* and *Protostrongylus stilesi* in Dall's sheep (*Ovis d. dalli*). Parasitol. 132:387–401. https://doi.org/10.1017/s0031182005009145.

Karns, P.D. 1967. *Pneumostrongylus tenuis* in deer in Minnesota and implications for moose. J. Wildl. Manage. 31:299–303. https://doi.org/10.2307/3798320.

Kutz, S.J., J. Ducrocq, G.G. Verocai, B.M. Hoar, D.D. Colwell, K.B. Beckmen, L. Polley, B.T. Elkin, and E.P. Hoberg. 2012. Parasites in ungulates of Arctic North America and Greenland: A view of contemporary diversity, ecology, and impact in a world under change. In: D. Rollinson and S.I. Hay (eds.), Advances in Parasitology, vol. 79, pp. 99–252. https://doi.org/10.1016/b978-0-12-398457-9.00002-0.

Kutz, S.J., E.P. Hoberg, and L. Polley. 2001. A new lungworm in muskoxen: An exploration in Arctic parasitology. Trends Parasitol. 17:276–80. https://doi.org/10.1016/s1471-4922(01)01882-7.

Lankester, M.W. 2001. Extrapulmonary lungworms of cervids. In: W.M. Samuel, M.J. Pybus, and A.A. Kocan (eds.), Parasitic diseases of wild mammals. 2nd ed., pp. 228–78. Ames (IA): Iowa State Univ. Press.

Prestwood, A.K. 1970. Neurologic disease in a white-tailed deer massively infected with meningeal worm (*Pneumostrongylus tenuis*). J. Wildl. Dis. 6:84–6. https://doi.org/10.7589/0090-3558-6.1.84.

Prestwood, A.K., V.F. Nettles, and F.E. Kellogg. 1974. Distribution of muscleworm, *Parelaphostrongylus andersoni*, among white-tailed deer of the southeastern United States. J. Wildl. Dis. 10:404–9. https://doi.org/10.7589/0090-3558-10.4.404.

Prestwood, A.K., and J.F. Smith. 1969. Distribution of meningeal worm (*Pneumostrongylus tenuis*) in deer in the southeastern United States. J. Parasitol. 55:720–5. https://doi.org/10.2307/3277200.

Pybus, M.J., W.M. Samuel, D. Welch, and C.J. Wilke. 1990. *Parelaphostrongylus andersoni* (Nematoda: Protostrongylidae) in white-tailed deer from Michigan. J. Wildl. Dis. 26:535–7. https://doi.org/10.7589/0090-3558-26.4.535.

Samuel, W.M., T.R. Platt, and S.M. Knispel-Krause. 1985. Gastropod intermediate hosts and transmission *of Parelaphostrongylus odocoilei*, a muscle-inhabiting nematode of mule deer, *Odocoileus h. hemionus*, in Jasper National Park, Alberta. Can. J. Zool. 63:928–32. https://doi.org/10.1139/z85-138.

Shostak, A.W., and W.M. Samuel. 1984. Moisture and temperature effects on survival and infectivity of first-stage larvae of *Parelaphostrongylus odocoilei* and *P. tenuis* (Nematoda: Metastrongyloidea). J. Parasitol. 70:261–9. https://doi.org/10.2307/3281873.

Slomke, A.M., M.W. Lankester, and W.J. Peterson. 1995. Infrapopulation dynamics of *Parelaphostrongylus tenuis* in white-tailed deer. J. Wildl. Dis. 31:125–35. https://doi.org/10.7589/0090-3558-31.2.125.

11. The Tale of the Tape: The World's Longest Parasite

Johnson, K.P., S.E. Bush, and D.H. Clayton. 2005. Correlated evolution of host and parasite body size: Tests of Harrison's Rule using birds and lice. Evol. 59:1744–53. https://doi.org/10.1111/j.0014-3820.2005.tb01823.x.

Kirk, W.D.J. 1991. The size relationship between insects and their hosts. Ecol. Entomol. 16:351–9. https://doi.org/10.1111/j.1365-2311.1991.tb00227.x.

Li, J., and E. Guo. 2016. *Taenia saginata* infestation. N. Engl. J. Med. 374:263. https://doi.org/10.1056/NEJMicm1504801.

Morand, S., M.S. Hafner, R.D.M. Page, and D.L. Reed. 2000. Comparative body size relationships in pocket gophers and their chewing lice. Biol. J. Linn. Soc. 70:239–49. https://doi.org/10.1111/j.1095-8312.2000.tb00209.x.

Morand, S., and R. Poulin. 2002. Body size–density relationships and species diversity in parasitic nematodes: Patterns and likely processes. Evol. Ecol. Res. 4:951–61. http://www.evolutionary-ecology.com/issues/v04n07/ddar1416.pdf.

Roberts, L.S., J.J. Janovy, Jr., and S. Nadler. 2013. Foundations of parasitology. 9th ed. New York: McGraw Hill.

Scholz, T., H.H. Garcia, R. Kuchta, and B. Wicht. 2009. Update on the human broad tapeworm (genus *Diphyllobothrium*), including clinical relevance. Clin. Microbiol. Rev. 22:146–60. https://doi.org/10.1128/CMR.00033-08.

Skrjabin, A.S. 1967. A gigantic diphyllobothriid *Polygonoporus giganticus* n.g., n. sp. – Sperm whale parasite [in Russian]. Parazitologiya 1:131–6.

Tsai, M-L., J-J. Li, and C-F. Dai. 2003. How host size may constrain the evolution of parasite body size and clutch size. The parasitic isopod *Icthyoxenus fushanensis* and its host fish, *Varicorhinus bacbatulus*, as an example. Oikos 92:13–19. https://doi.org/10.1034/j.1600-0706.2001.920102.x.

Voris, H.K., W.B. Jeffries, and S. Poovachiranon. 2000. Size and location relationships of stalked barnacles of the genus *Octolasmis* on the mangrove crab *Scylla serrata*. J. Crust. Biol. 20:483–94. https://doi.org/10.1163/20021975-99990064.

12. Death by Raccoon

Cox, F.E.G. 2002. History of human parasitology. Clin. Microbiol. Rev. 15:595–612. https://doi.org/10.1128/CMR.15.4.595-612.2002.

Fox, A.S., K.R. Kazacos, N.S. Gould, P.T. Heydemann, C. Thomas, and K.M. Boyer. 1985. Fatal eosinophilic meningoencephalitis and visceral larva migrans caused by the raccoon ascarid *Baylisascaris procyonis*. N. Engl. J. Med. 312:1619–23. https://doi.org/10.1056/nejm198506203122507.

Grassi, B. 1888. Weiteres zur frage der Ascarisentwickelung [More on the question of Ascaris development]. Centralblatt für Bakter. und Parasitenkunde 3:748–9.

Huff, D.S., R.C. Neafie, M.J. Binder, G.A. De Leon, L.W. Brown, and K.R. Kazacos. 1984. Case 4. The first fatal *Baylisascaris* infection in humans: An infant with eosinophilic meningoencephalitis. Ped. Path. 2: 345–52. https://doi.org/10.3109/15513818409022268.

Kazacos, K.R. 2001. *Baylisascaris procyonis* and related species. In: W.M. Samuel, M.J. Pybus, and A.A. Kocan (eds.), Parasitic diseases of wild mammals. 2nd ed., pp. 302–41. Ames (IA): Iowa State Univ. Press.

Polley, L., and A. Thompson. 2015. Parasites and wildlife in a changing world. Trends Parasitol. 31:123–4. https://doi.org/10.1016/j.pt.2015.03.001.

Rainwater, K.L., K. Marchese, S. Slavinski, L.E. Humberg, E.J. Dubovi, J.A. Jarvis, D. McAloose, and P.P. Calle. 2017. Health survey of free-ranging raccoons (*Procyon lotor*) in Central Park, New York, New York, USA: Implications for human and domestic animal health. J. Wildl. Dis. 53:272–84. https://doi.org/10.7589/2016-05-096.

Sorvillo, F., L.R. Ash, O.G.W. Berlin, J. Yatabe, C. Degiorgio, and S.A. Morse. 2002. *Baylisascaris procyonis*: An emerging helminthic zoonosis. Emerg. Infect. Dis. 8:355–9.

Weinstein, S.B. 2016. *Baylisascaris procyonis* demography and egg production in a California raccoon population. J. Parasitol. 102:622–8. https://doi.org/10.1645/15-747.

Weinstein, S.B., and K.D. Lafferty. 2015. How do humans affect wildlife nematodes? Trends Parasitol. 31:222–7. https://doi.org/10.1016/j.pt.2015.01.005.

Weinstein, S.B., C.M. Lake, H.M. Chastain, D. Fisk, S. Handali, P.L. Khan, S.P. Montgomery, P.P. Wilkins, A.M. Kuris, and K.D. Lafferty. 2017. Seroprevalence of *Baylisascaris procyonis* infection among humans, Santa Barbara County, California, USA, 2014–2016. Emerg. Infect. Dis. 23:1397–19. https://doi.org/10.3201/eid2308.170222.

Weinstein, S.B., C.W. Moura, J.F. Mendez, and K.D. Lafferty. 2018. Fear of feces? Tradeoffs between disease risk and foraging drive animal activity around raccoon latrines. Oikos 127:927–34. https://doi.org/10.1111/oik.04866.

Wise, M.E., F.J. Sorvillo, S.C. Shafir, L.A. Ash, and O.G. Berlin. 2005. Severe and fatal central nervous system disease in humans caused by *Baylisascaris procyonis*, the common roundworm of raccoons: A review of current literature. Microbes Infect. 7:317–23. https://doi.org/10.1016/j.micinf.2004.12.005.

13. Moths, Sloths, Tears, and Blood

Bänziger, H. 1989. Skin-piercing blood-sucking moths V: Attacks on man by 5 *Calyptra* spp. (Lepidoptera, Noctuidae) in S and SE Asia. Mitt. Schweiz Ent. Ges. 62:215–33.

Büttiker, W., H.W. Krenn, and J.F. Putterill. 1996. The proboscis of eye-frequenting and piercing Lepidoptera (Insecta). Zoomorphology 116:77–83. https://doi.org/10.1007/BF0252687.

Hill, S.R., J. Zaspel, S. Weller, B.S. Hansson, and R. Ignell. 2010. To be or not to be…a vampire: A matter of sensillum numbers in *Calyptra thalictra*? Arthropod Struct. Dev. 39:322–33. https://doi.org/10.1016/j.asd.2010.05.005.

Kircher, J. 1997. A neotropical companion: An introduction to the animals, plants and ecosystems of the new world tropics. 2nd ed. Princeton (NJ): Princeton Univ. Press.

Pauli, J.N., J.E. Mendoza, S.A. Steffan, C.C. Carey, P.J. Weimer, and M.Z. Peery. 2014. A syndrome of mutualism reinforces the lifestyle of a sloth. Proc. Roy. Soc. Lond. B. 281:20133006. https://doi.org/10.1098/rspb.2013.3006.

Plotkin, D., and J. Goddard. 2013. Blood, sweat and tears: A review of the hematophagous, sudophagous, and lachryphagous Lepidoptera. J. Vector Ecol. 38:289–94. https://doi.org/10.1111/j.1948-7134.2013.12042.x.

Sibaja-Morales, K.D., J.B. de Olivera, A.E. Jiménez Rocha, J. Hernández Gamboa, J. Prendas Gamboa, F. Arroyo Murillo, J. Sandí, Y. Nuñez, and M. Baldi. 2009. Gastrointestinal parasites and ectoparasites of *Bradypus variegatus* and *Choloepus hoffmanni* sloths in captivity from Costa Rica. J. Zoo Wildl. Med. 40:86–90. https://doi.org/10.1638/2008-0036.1.

Waage, J.K. 1979. The evolution of insect/vertebrate associations. Biol. J. Linn. Soc. 12:187–224. https://doi.org/10.1111/j.1095-8312.1979.tb00055.x.

Waage, J.K., and G.G. Montgomery. 1976. *Cryptoses choloepi*: A coprophagous moth that lives on a sloth. Science 193:157–8. https://doi.org/10.1126/science.193.4248.157.

Zaspel, J.M., S.J. Weller, and M.A. Branham. 2011. A comparative survey of the proboscis morphology and associated structures in fruit-piercing, tear-feeding, and blood-feeding moths in Calpinae (Lepidoptera: Erebidae). Zoomorphology 130:203–25. https://doi.org/10.1007/s00435-011-0132-1.

Zaspel, J.M., R. Zahiri, M.A. Hoy, D. Janzen, S.J. Weller, and N. Wahlberg. 2012. A molecular phylogenetic analysis of the vampire moths and their fruit-piercing relatives (Lepidoptera: Erebidae: Calpinae). Mol. Phylogenet. Evol. 65:786–91. https://doi.org/10.1016/j.ympev.2012.06.029.

14. The Manchurian Parasite

Abele, L.G., and S. Gilchrist. 1977. Homosexual rape and sexual selection in acanthocephalan worms. Science 197:81–3. https://doi.org/10.1126/science.867055.

Berdoy, M., J.P. Webster, and W.M. Macdonald. 1995. Parasite-altered behavior: Is the effect of *Toxoplasma gondii* on *Rattus norvegicus* specific? Parasitology 111: 403–9. https://doi.org/10.1017/s0031182000065902.

Bethel, W.M., and J.C. Holmes. 1973. Altered evasive behavior and responses to light in amphipods harboring acanthocephalan cystacanths. J. Parasitol. 59:945–56. https://doi.org/10.2307/3278623.

Bowen, R.C. 1967. Defense reactions of certain spirobolid millipedes to larval *Macracanthorhynchus ingens*. J. Parasitol. 53:1092–5. https://doi.org/10.2307/3276846.

Crompton, D.W.T. 1970. An ecological approach to acanthocephalan physiology. Cambridge: Cambridge Univ. Press.

Helluy, S. 2013. Parasite-induced alterations of sensorimotor pathways in gammarids: Collateral damage of neuroinflammation? J. Exp. Biol. 216:67–77. https://doi.org/10.1242/jeb.073213.

Herlyn, H., O. Piskurek, J. Schmitz, U. Ehlers, and H. Zischler. 2003. The syndermatan phylogeny and the evolution of acanthocephalan endoparasitism as inferred from 18S rDNA sequences. Mol. Phylogenet. Evol. 26:155–64. https://doi.org/10.1016/s1055-7903(02)00309-3.

Holmes, J.C., and W.M. Bethel. 1972. Modifications of intermediate host behaviour by parasites. In: E.U. Canning and C.A. Wright (eds.), Behavioural aspects of parasite transmission, pp. 123–49. Suppl. 1 of Zoo. J. Lin. Soc. 51. London: Academic Press.

Kennedy, C.R. 2006. Ecology of the acanthocephala. Cambridge: Cambridge Univ. Press.

Lefèvre, T., S.A. Adamo, D.G. Biron, D. Missé, D. Hughs, and F. Thomas. 2009. Invasion of the body snatcher: The diversity and evolution of manipulative strategies in host–parasite interactions. Adv. Parasitol. 68:45–83. https://doi.org/10.1016/s0065-308x(08)00603-9.

Libersat, F., and J. Moore. 2000. The parasite *Moniliformis moniliformis* alters the escape response of its cockroach host *Periplaneta americana*. J. Insect Behav. 13:103–10. https://doi.org/10.1023/A:1007719710664.

McAuliffe, K. 2017. This is your brain on parasites: How tiny creatures manipulate our behavior and shape society. Boston: Mariner Books.

Moore, J. 1983. Altered behavior in cockroaches (*Periplaneta americana*) infected with an archaeacanthocephalan, *Moniliformis moniliformis*. J. Parasitol. 69:1774–6.

Moore, J. 2002. Parasites and the behavior of animals. New York: Oxford Univ. Press.

Moore, J., and M. Freehling. 2002. Cockroach hosts in thermal gradients suppress parasite development. Oecologia 133:261–6. https://doi.org/10.1007/s00442-002-1030-5.

Servick, K. 2020. Brain parasite may strip away rodents' fear of predators – not just of cats. Science, January 14, 2020. https://www.science.org/content/article/brain-parasite-may-strip-away-rodents-fear-predators-not-just-cats.

Sinisalo, T., R. Poulin, H. Högmander, T. Juuti, and E.T. Valtonen. 2004. The impact of sexual selection on *Corynosoma magdaleni* (Acanthocephala) infrapopulations in Saimaa ringed seals (*Phoca hispida saimensis*). Parasitology 128:179–85. https://doi.org/10.1017/s003118200300430x.

Sørensen, M.V., and G. Giribet. 2006. A modern approach to rotiferan phylogeny: Combining morphological and molecular data. Mol. Phylogenet. Evol. 40:585–608. https://doi.org/10.1016/j.ympev.2006.04.001.

Vyas, A. 2015. Mechanisms of host behavioral change in *Toxoplasma gondii* rodent association. PLoS Pathog. 11:e1004935. https://doi.org/10.1371/journal.ppat.1004935.

Webster, J.P. 2007. The effect of *Toxoplasma gondii* on animal behavior: Playing cat and mouse. Schizophr. Bull. 33:752–6. https://doi.org/10.1093/schbul/sbl073.

Wilson, K., and J. Edwards. 1986. The effects of parasitic infection on the behaviour of an intermediate host, the American cockroach, *Periplaneta americana*, infected with the acanthocephalan, *Moniliformis moniliformis*. Anim. Behav. 34:942–4. https://doi.org/10.1016/S0003-3472(86)80088-4.

15. A Ghost of a Chance

Addison, E.M., and R.F. McLaughlin. 1988. Growth and development of winter tick, *Dermacentor albipictus*, on moose, *Alces alces*. J. Parasitol. 74:670–8. https://doi.org/10.2307/3282188.

Addison, E.M., R.D. Strickland, and D.J.H. Fraser. 1989. Gray jays, *Perisoreus canadensis*, and common ravens, *Corvus corax*, as predators of winter ticks, *Dermacentor albipictus*. Can. Field Nat. 103:406–8.

Allan, S.A. 2001. Ticks (class Arachnida: order Acarina). In: W.M. Samuel, M.J. Pybus, and A.A. Kocan (eds.), Parasitic diseases of wild mammals. 2nd ed., pp. 72–106. Ames (IA): Iowa State Univ. Press.

Banks, B. 2017. The mystery of our disappearing moose. Can. Wildl. 23:18–26.

Drew, M.L., and W.M. Samuel. 1985. Factors affecting transmission of larval winter ticks, *Dermacentor albipictus* (Packard), to moose, *Alces alces* L., in Alberta, Canada. J. Wildl. Dis. 21:274–82. https://doi.org/10.7589/0090-3558-21.3.274.

Drew, M.L., and W.M. Samuel. 1986. Reproduction of the winter tick, *Dermacentor albipictus*, under field conditions in Alberta, Canada. Can. J. Zool. 64:714–21. https://doi.org/10.1139/z86-105.

Dybas, C.L. 2009. Minnesota's moose: Ghosts of the northern forest? BioSci. 59:824–8. https://doi.org/10.1525/bio.2009.59.10.3.

Glines, M.V., and W.M. Samuel. 1989. Effect of *Dermacentor albipictus* (Acari: Ixodidae) on blood composition, weight gain and hair coat of moose, *Alces alces*. Experiment. Appl. Acarol. 6:197–213. https://doi.org/10.1007/bf01193980.

Holmes, C.J., C.D. Dobrotka, D.W. Farrow, A.J. Rosendale, J.B. Benoit, P.J. Pekins, and J.A. Yoder. 2018. Low and high thermal tolerance characteristics for unfed larvae of the winter tick *Dermacentor albipictus* (Acari: Ixodidae) with special reference to moose. Ticks Tick Borne Dis. 9:25–30. https://doi.org/10.1016/j.ttbdis.2017.10.013.

Killick, A. 1999. Unusually high number of ticks may kill about 1000 Manitoba moose: Itching drives them crazy. National Post, April 27.

McLaughlin, R.F., and E.M. Addison. 1986. Tick (*Dermacentor albipictus*) – induced winter hair-loss in captive moose (*Alces alces*). J. Wildl. Dis. 22:501–10. https://doi.org/10.7589/0090-3558-22.4.502.

McPherson, M., A.W. Shostak, and W.M. Samuel. 2000. Climbing simulated vegetation to heights of ungulate hosts by larvae of *Dermacentor albipictus* (Acari: Ixodidae). J. Med. Entomol. 37:114–20. https://doi.org/10.1603/0022-2585-37.1.114.

Raup, D.M. 1991. Extinction: Bad genes or bad luck? London: Norton.

Rosen, Y. 2017. Threat of moose-killing tick infestation looms as far-north climate warms. Alaska Dispatch News. December 2, pp. 1–5.

Samuel, B. 2004. White as a ghost: Winter ticks and moose. Edmonton (AB): Federation of Alberta Naturalists.

Samuel, W.M. 2007. Factors affecting epizootics of winter ticks and mortality of moose. Alces 43:39–48. https://alcesjournal.org/index.php/alces/article/view/349.

Samuel, W.M., and D.A. Welch. 1991. Winter ticks on moose and other ungulates: Factors influencing their population size. Alces 27:169–82. https://www.alcesjournal.org/index.php/alces/article/view/1119.

Welch, D.A., W.M. Samuel, and C.J. Wilke. 1991. Suitability of moose, elk, mule deer, and white-tailed deer as hosts for winter ticks (*Dermacentor albipictus*). Can. J. Zool. 69:2300–5. https://doi.org/10.1139/z91-323.

16. Sex and the Single Schistosome

Brant, S.V., and E.S. Loker. 2005. Can specialized pathogens colonize distantly related hosts? Schistosome evolution as a case study. PLoS Pathog. 1:e38. https://doi.org/10.1371/journal.ppat.0010038.

Brown, S.P., and B.T. Grenfell. 2001. An unlikely partnership: Parasites, concomitant immunity and host defense. Proc. R. Soc. Lond. B. 268:2543–9. https://doi.org/10.1098/rspb.2001.1821.

Després, L., D. Imbert-Establet, C. Combes, and F. Bonhomme. 1992. Molecular evidence linking hominid evolution to recent radiation of schistosomes (Platyhelminthes: Trematoda). Mol. Phylogenet. Evol. 1:295–304. https://doi.org/10.1016/1055-7903(92)90005-2.

Després, M., and S. Maurice. 1995. The evolution of dimorphism and separate sexes in schistosomes. Proc. R. Soc. Lond. B. 262:175–80. https://www.jstor.org/stable/50214.

Kuperschmidt, K. 2018. Worms living in your veins? Seventeen volunteers said "OK." Science 359:853–4. https://doi.org/10.1126/science.359.6378.853.

Kusel, J.R., B.H. Al-Adhami, and M.J. Doenhoff. 2007. The schistosome in the mammalian host: Understanding the mechanisms of adaptation. Parasitology 134:1477–1526. https://doi.org/10.1017/s0031182007002971.

Lawton, S.P., H. Hirai, J.E. Ironside, D.A. Johnston, and D. Rollinson. 2011. Genomes and geography: Genomic insights into the evolution and phylogeography of the genus *Schistosoma*. Parasit. Vectors 4:131. https://doi.org/10.1186/1756-3305-4-131.

LoVerde, P.T., E.G. Niles, A. Osman, and W. Wu. 2004. *Schistosoma mansoni* male-female interactions. Can. J. Zool. 82:357–74. https://doi.org/10.1139/z03-217.

Lu, Z., F. Sessler, N. Holroyd, S. Hahnel, T. Quack, M. Berriman, and C.G. Grevelding. 2016. Schistosome sex matters: A deep view into gonad-specific and pairing-dependent transcriptomes reveals a

complex gender interplay. Sci. Rep. 6:31150. https://doi.org/10.1038
/srep31150.
Morand, S., and C.D. Müller-Graf. 2000. Muscles or testes? Comparative
evidence for sexual competition among dioecious blood parasites
(Schistosomatidae) of vertebrates. Parasitology 120:45–56. https://
doi.org/10.1017/s0031182099005235.
Platt, T.R., and D.R. Brooks. 1997. Evolution of the schistosomes
(Digenea: Schistosomatoidea): The origin of dioecy and colonization
of the venous system. J. Parasitol. 83:1035–44. https://doi.org
/10.2307/3284358.
Rauch, G., M. Kalbe, and T.B.H. Reusch. 2005. How a complex life cycle
can improve a parasite's sex life. J. Evol. Biol. 18:1069–75. https://doi
.org/10.1111/j.1420-9101.2005.00895.x.
Webster, J.P., J.I. Hoffman, and M. Berdoy. 2003. Parasite infection, host
resistance and mate choice: Battle of the genders in a simultaneous
hermaphrodite. Proc. Roy. Soc. B. 270:1481–5. https://doi.org/10.1098
/rspb.2003.2354.
World Health Organization (WHO). 2017. Guideline on control and
elimination of human schistosomiasis. Geneva: WHO.
World Health Organization (WHO). 2023. Fact sheet: Schistosomiasis.
Geneva: WHO. https://www.who.int/news-room/fact-sheets/detail
/schistosomiasis.
Zhang, G., O. Verneau, C. Qiu, J. Jourdane, and M. Xia. 2001. An African
or Asian evolutionary origin for human schistosomes? C.R. Acad. Sci.
Paris. 324:1001–10. https://doi.org/10.1016/s0764-4469(01)01383-x.

17. The Trickster: Coyotes and Their Parasites

Borstein, S., T. Mörner, and W.M. Samuel. 2001. *Sarcoptes scabiei* and
sarcoptic mange. In: W.M. Samuel, M.J. Pybus, and A.A. Kocan (eds.),
Parasitic diseases of wild mammals. 2nd ed., pp. 107–19. Ames (IA):
Iowa State Univ. Press.
Bradley, C.A., and S. Altizer. 2007. Urbanization and the ecology of
wildlife diseases. Trends Ecol. Evol. 22:95–102. https://doi.org
/10.1016/j.tree.2006.11.001.
City of Edmonton. n.d. The Edmonton Urban Coyote Project. https://
edmontonurbancoyotes.ca/.
Flores, D. 2016. Coyote America: A natural and supernatural history.
New York: Basic Books.

Gesy, K.M., J.M. Schurer, A. Massolo, S. Liccioli, B.T. Elkin, R. Alisaukas, and E.J. Jenkins. 2014. Unexpected diversity of the cestode *Echinococcus multilocularis* in wildlife in Canada. Intl. J. Parasitol. Parasites Wildl. 3:81–7. https://doi.org/10.1016/j.ijppaw.2014.03.002.

Grinder, M., and P.R. Krausman. 2001. Morbidity-mortality factors and survival of an urban coyote population in Arizona. J. Wildl. Dis. 37:312–17. https://doi.org/10.7589/0090-3558-37.2.312.

Holmes, J.C., J.L. Mahrt, and W.M. Samuel. 1971. The occurrence of Echinococcus multilocularis Leuckart, 1863 in Alberta. Can. J. Zool. 49:575–6. https://doi.org/10.1139/z71-090.

Holmes, J.C., and R. Podesta. 1968. The helminths of wolves and coyotes from the forested regions of Alberta. Can. J. Zool. 46:1193–1204. https://doi.org/10.1139/z68-169.

Houston, S., S. Belga, K. Buttenschoen, R. Cooper, S. Girgis, B. Gottstein, G. Low, A. Massolo, C. MacDonald, N. Müller, J. Preiksaitis, P. Sarlieve, S. Vaughn, and K. Kowalewska-Grochowska. 2021. Epidemiological and clinical characteristics of alveolar echinococcosis: An emerging infectious disease in Alberta, Canada. Am. J. Trop. Med. Hyg. 104:1863–9. https://doi.org/10.4269/ajtmh.20-1577.

Liccioli, S., S. Catalano, S.J. Kutz, M. Lejeune, G.G. Verocai, P.J. Duignan, C. Fuentealba, M. Hart, K.E. Ruckstuhl, and A. Massolo. 2012. Gastrointestinal parasites of coyotes (*Canis latrans*) in the metropolitan area of Calgary, Alberta, Canada. Can. J. Zool. 90:1023–30. https://doi.org/10.1139/z2012-070.

Luong, L.T., J.L. Chambers, A. Moizis, T.M. Stock, and C.C. St. Clair. 2020. Helminth parasites and zoonotic risk associated with urban coyotes (*Canis latrans*) in Alberta, Canada. J. Helminthol. 94:e25. https://doi.org/10.1017/S0022149X1800113X.

Massolo, A., S. Liccioli, C. Budke, and C. Klein. 2014. *Echinococcus multilocularis* in North America: The great unknown. Parasite 21:73. https://doi.org/10.1051/parasite/2014069.

Murray, M.H., J. Hill, P. Whyte, and C. C. St. Clair. 2016. Urban compost attracts coyotes, contains toxins, and may promote disease in urban-adapted wildlife. EcoHealth 13:285–92. https://doi.org/10.1007/s10393-016-1105-0.

Samuel, W.M., S. Ramalingam, and L.N. Carbyn. 1978. Helminths in coyotes (*Canis latrans* Say), wolves (*Canis lupus* L.) and red foxes (*Vulpes vulpes* L.) of southwestern Manitoba. Can. J. Zool. 56:2614–17. https://doi.org/10.1139/z78-351.

Schilthuizen, M. 2018. Darwin comes to town: How the urban jungle drives evolution. New York: Picador Books.

Statham, M.J., B.N. Sachs, K.B. Aubry, J.D. Perrine, and S.M. Wisely. 2012. The origin of recently established red fox populations in the United States: Translocations or natural range expansions? J. Mammal. 93:52–65. https://doi.org/10.1644/11-MAMM-A-033.1.

18. Fleas: The Inside Story

Bitam, I., K. Dittmar, P. Parola, M.F. Whiting, and D. Raoult. 2010. Fleas and flea-borne diseases. Int. J. Infect. Dis. 14:e667–76. https://doi.org/10.1016/j.ijid.2009.11.011.

De Morgan, A. 1915. A budget of paradoxes. vol. II. London: Open Court Publ.

Dittmar, K., Q. Zhu, M.W. Hastriter, and M.F. Whiting. 2016. On the probability of dinosaur fleas. BMC Evol. Biol. 16:9. https://doi.org/10.1186/s12862-015-0568-x.

Hooke, R. 1665. Micrographia: Or, some physiological descriptions of minute bodies made by magnifying glasses with observations and descriptions thereupon. London: Jo. Martyn and Ja. Allestry.

Krasnov, B.R. 2008. Functional and evolutionary ecology of fleas: A model for ecological parasitology. Cambridge: Cambridge Univ. Press.

Linardi, P.M., J-C Beaucournu, D. Moreira de Avelar, and S. Belaz. 2014. Notes on the genus *Tunga* (Siphonaptera: Tungidae) II – neosomes, morphology, classification, and other taxonomic notes. Parasite 21:68 https://doi.org/10.1051/parasite/2014067.

Linardi, P.M., and D. Moreira de Avelar. 2014. Neosomes of tungid fleas on wild and domestic animals. Parasitol. Res. 113:3517–33. https://doi.org/10.1007/s00436-014-4081-8.

Marshall, A.G. 1981. The ecology of ectoparasitic insects. London: Academic Press.

Martin, A.J. 2017. The evolution underground: Burrows, bunkers, and the marvelous subterranean world beneath our feet. New York: Pegasus Books.

Stewart, A. 2011. Wicked bugs: The louse that conquered Napoleon's army and other diabolical insects. Chapel Hill (NC): Algonquin Books.

Tomassini, R.L., C.I. Montalvo, and M.C. Ezquiaga. 2016. The oldest record of flea/armadillos interaction as example of bioerosion on osteoderms from the late Miocene of the Argentine pampas. Int. J. Paleopathol. 15:65–8. https://doi.org/10.1016/j.ijpp.2016.08.004.

Traub, R. 1980. The zoogeography and evolution of some fleas, lice and mammals. In: H. Starcke and R. Traub (eds.), Fleas. Rotterdam: A.A. Balkema.

Whiting, M.F. 2002. Mecoptera is paraphyletic: Multiple genes and phylogeny of Mecoptera and Siphonaptera. Zool. Scr. 31:93–104. https://doi.org/10.1046/j.0300-3256.2001.00095.x.

Whiting, M.F., A.S. Whiting, M.W. Hastriter, and K. Dittmar. 2008. A molecular phylogeny of fleas (Insecta: Siphonaptera): Origins and host associations. Cladistics 24:677–707. https://doi.org/10.1111/j.1096-0031.2008.00211.x.

Zhu, Q., M.H. Hastriter, M.F. Whiting, and K. Dittmar. 2015. Fleas (Siphonaptera) are Cretaceous, and evolved with Theria. Mol. Phylogenet. Evol. 90:129–39. https://doi.org/10.1016/j.ympev.2015.04.027.

Conclusion: The Greatest Show on Earth

Combes, C. 2005. The art of being a parasite. (Translated by D. Simberloff). Chicago: Univ. Chicago Press.

Dawkins, R. 2009. The greatest show on earth: The evidence for evolution. New York: Free Press.

Dogiel, V.A. 1964. General parasitology. London: Oliver and Boyd.

Goater, T.M., C.P. Goater, and G.W. Esch. 2014. Parasitism: The diversity and ecology of animal parasites. 2nd ed. Cambridge: Cambridge Univ. Press.

Gordon, C.A., D.P. McManus, M.K. Jones, D.J. Gray, and G.N. Gobert. 2016. The increase of exotic zoonotic helminth infections: The impact of urbanization, climate change and globalization. Adv. Parasitol. 91:311–97. https://doi.org/10.1016/bs.apar.2015.12.002.

Jokela, J., M.F. Dybdahl, and C.M. Lively. 2009. The maintenance of sex, clonal dynamics, and host-parasite coevolution in a mixed population of sexual and asexual snails. Am. Nat. 174(Suppl 1):S43–53. https://doi.org/10.1086/599080.

Lively, C.M. 1987. Evidence from a New Zealand snail for the maintenance of sex by parasitism. Nature 328:519–21. https://doi.org/10.1038/328519a0.

Loker, E.S., and B.V. Hofkin. 2015. Parasitology: A conceptual approach. New York: Garland Science.

Margulis, L. 1981. Symbiosis in cell evolution. San Francisco: W.H. Freeman.

Marshall, A.G. 1981. The ecology of ectoparasitic insects. London: Academic Press.

Miller, M.R., A. White, and M. Boots. 2006. The evolution of parasites in response to tolerance in their hosts: The good, the bad, and apparent commensalism. Evolution 60:945–56. https://doi.org/10.1554/05-654.1.

Moore, J. 2002. Parasites and the behavior of animals. Oxford: Oxford Univ. Press.

Polley, L., and A. Thompson. 2015. Parasites and wildlife in a changing world. Trends Parasitol. 31:123–4. https://doi.org/10.1016/j.pt.2015.03.001.

Roberts, L.S., and J. Janovy, Jr. 2009. Gerald D. Schmidt and Larry S. Roberts' foundations of parasitology. 8th ed. Boston: McGraw Hill.

Samuel, W.M., M.J. Pybus, and A.A. Kocan (eds.). 2001. Parasitic diseases of wild mammals. 2nd ed. Ames (IA): Iowa State Univ. Press.

Index

Page numbers in italics represent illustrations.

aardvarks, 186
acanthella, 179
acanthocephalans, 176, 178–84, 186–7. *See also* altered behaviour; mind-control by parasites; thorny-headed worms
acanthors, 179
adaptive immunity, 110
Addison's disease, 259
Africa: bushbabies/galagos, 15; giant deer, 97; Grant's gazelle, 109; ground squirrels, 15–18; human lice, 39; lemurs, 17; pinworms, 15–18; porcupines, 19; primate evolution, 15–16, 18; schistosomiasis, 216–17; *Trichinella* roundworms, 28; tsetse flies, 89; vampire moths, 164; waterholes, 205; zebras, 83, 86
African buffalo, 103–4

African elephants, 105
African sleeping sickness, 89
air sinuses, 75, 76, 79
Alaria americana, 234. *See also* flatworms
Alaria arisaemoides, 234. *See also* flatworms
Alaska, 27, 29, 52, 131, 137
Alberta: caribou, 136; coyotes, 228–35; giant liver flukes, 52; hydatid disease, 230; moose, 198; muscle worms, 132; research on scuds, 179–80
Alberta, University of, 233
Alces alces, 194. *See also* moose
algae, 171
allergic response, 243
allogrooming, 41, 116, 120, 122, 260
altered behaviour, 176–8, 179–84, 190

American beavers, 61
amphibians, 46
amphipods, 180, 181, 184, 187
anacondas, 170
Anoplura, 34. *See also* lice
Antarctica, 144, 145
anthrax, 206
anthropogenic foods, 235
antibodies, 208, 209
anticoagulants, 164, 243
antilocaprids, 100, 164
antlers, 96–8, 100, 108
ants, 189–90
apes: great apes, 17, 34, 41; human evolution from, 38, 41; and lice, 32, 34, 44
Aplodontia, 61
arboreal habitat, 16, 170, 253
Arctic: climate change, 21, 28–9, 138, 264; muskoxen, 137; porcupines, 19; sperm whales, 145; *Trichinella* roundworms, 25–30; weather station in, 22
Arctic foxes, 232
Argentina, 115
Arizona, 228
armadillo, 252, *254*
Ascarids, 146, 150–3, 264. *See also* roundworms
Ascaris lumbricoides, 150–2
Ascaris summ, 150
Asia: evolution of primates, 15, 39–40; moths, 164; origin of roundworms, 28; plague bacteria, 32; porcupines, 19; red foxes, 232; schistosomiasis, 216–17
Asian elephants, 105
Asian guar, 100–1
asthma, 259
Atlantic Ocean, 18
Australia, 28, 60, 61, 249, 252
autoimmune diseases, 259

baboons, 15, 82–3
Babyrousa celebensis, 111f
Bacillus anthracis, 206
bacteria: as ancient parasites, 255–6; and body lice, 36–7; and consumption/tuberculosis, 114; and horse flies, 91; and lice, 31; and parasites, 7; and plague/Black Death, 32–4, 37, 243; and sloths, 169; and vampire bats, 123–4. *See also* anthrax; *Bartonella quintana*; *Yersinia pestis*
badgers, 153
Baltic region, 246
bandicoots, 61
barnacles, 142
Bartonella quintana, 37, 121, 122. *See also* trench fever
bat bugs, 116, 120–2, 124, 264
bat flies, 246, *247*
bats, 125
Baylisascaris procyonis, 153, 155–9, 229
bears, 143, 146, 153, 264. *See also* polar bears
beaver beetles: and beaver lodges, 65–6; behaviour of, 64–5; characteristics of, 62–4; as non-invasive species, 67; specializations of, 61–2, 68–9
beaver mites, 68
beavers: habitat of, 68; as hosts of beetles, 61–9; in Scotland, 67; and thorny-headed worms, 180, 186
beetles: diversity of, 58–9, 69; features of, 59; in fur of rodents and mammals, 60–1; in fur of sloths, 60; mountain pine beetles, 136, 202, 203; round fungus beetles, 61; rove beetles, 60–1. *See also* beaver beetles
Bergmann, Carl, 139, 147

Bergmann's rule, 139, 147
bighorn sheep, 101
biodiversity, 4, 57, 69, 237, 263
biology, science of, 4, 139, 142, 147, 256, 263
biomonitors, 178
birds: and bat bugs, 121; and *Baylisascaris* roundworms, 156; and Bergman's rule, 139; and blood flukes, 206–7, 215–16; colors of, 81; feeding on ticks, 200; and fleas, 245, 249, 254; and Hamilton-Zuk Hypothesis, 99; and lice, 239; and liver flukes, 46; and raccoon latrines, 264; and tapeworms, 146, 230; tears of, 167; and thorny-headed worms, 176; and urbanization, 221; and vampire bats, 115–17
bison, 100
Bison latifrons, 100
Black Death, 32–4, 37, 44, 240, 243. See also *Yersinia pestis*
black-tailed deer, 52, 132
blood flukes: in blood vessels, 208, 210; characteristics of, 206–7; effects on hosts, 214; eradication of, 219; evolution of, 215–16; and host defenses, 208–9, 214, 259, 262; and human infections, 216–17; life cycle of, 210–13; reproduction of, 206, *207*, 212–13, 217–19, 262; vaccine for, 220. See also schistosomiasis
body condition, 104, 107–8
body lice. See *under* lice
boreal forests, 194, 202, 203
Boreidae, 145
Bos gaurus, 100–1
bot flies, 88, 103
bothria, 144
bovids, 100, 103, 133, 167

Bradypodidae, 145, 168, 170
Bradypus, 81. See also sloths
brain worms, 128–30, 133, 134, 137. See also moose sickness
brains – animals: and flatworm flukes, 76–7, 78–9. See also altered behaviour; brain worms; mass stranding events; moose sickness; white-tailed deer
brains – humans: brainwashing, 176; children's brain infections, 158; and lice, 32; and *Trichinella* infection, 26
Brazil, 34, 251
breeding hierarchies, 107
British Columbia, 52, 136, 201, 202
buffalo, 103–4
bushbabies, 15, 17

caimans, 170
calcification, 25
Calgary, 228–31
California, 52, 115, 158
Calyptra, 163–6. See also moths
Cameron, T.W., 13–17
camouflage, 85–7, 171, 180, 209
Canada: brain worms in deer, 133; climate change, 55; coyote parasites, 236, 237; muskoxen, 137; raccoons, 153, 159; wolves and moose, 1–2
canis edwardii, 222. See also wolves
canis latrans, 222. See also coyotes
canis lupus, 222. See also wolves
Cape Breton Island, 201
Capra pyrenaica hispanica, 105
capybaras, 61, 120
Caribbean, 217, 254
caribou: antlers, 100, 103; barren ground caribou, 131, 265; and brain worms, 134, 136–7;

caribou (*continued*)
  declining populations of, 55, 127, 136–8, 265; and giant liver flukes, 52, 55; and muscle worms, 131–2; risk of extinction of, 265; woodland caribou, 132, 136
Carmilla (Le Fanu), 114
carnivores, 26–8, 117, 124, 150, 153, 207, 244
caruncles, 225
Castor canadensis, 61. See also beavers
Castor fiber, 61, 67. See also beavers
cats: domestic, 12, 153, 238; wild ("big cats"), 188–9. See also *Toxoplasma gondii*
cattle, 54, 89, 93–4, 100
caves, 20, 115
Cecropia, 162
cement, 34, 178, 186, 196
cement glands, 178, 186
Central America, 19, 153, 169
central nervous system, 26, 76, 77, 130, 134
cephalopods, 73, 145
Ceratophyllus lunatus, 245
cercaria, 49–50, 211
cerebrospinal nematodiasis, 128. See also moose sickness
Cervalces scotti, 97. See also moose
cervids, 54, 133, 135, 231
Cervus elaphus elaphus, 53. See also deer
Cestodes, 46
cetaceans, 71–2, 75. See also dolphins; whales
chelicarae, 196
chigoes, 250
children, infections in, 10, 13, 157–8
chimpanzees, 14, 15, 34, 41–2
China, 19
cholesterol, 51
Choloepus, 81. See also sloths

chromosomes, 10–12, 123, 261
chronic effects, 236
cladistic analysis, 16
cladistics, 14, 19
clasper, 243
Clethrionomys gapperi, 233
climate/environmental change: in the Arctic, 21, 28–9, 138; and brain worms, 127; caribou habitat, 136–7; general effects of, 8, 258–9, 263–5; and giant liver flukes, 54–6; and marine life, 71–2, 79–80; and moose, 135, 136, 202–3; and winter ticks, 202–3
coccideans, 104
cockroaches, 182–5
coevolution, 13, 19, 43, 107, 134, 237, 263
Coleoptera, 59. See also beetles
commensals, 60, 67–8
complement, 208
compost sites, 236
concomitant immunity, 214, 218
Condon, Richard, 176
convergent evolution, 177, 188–9
coprolites, 20, 151
copulatory bursa, 178
coronavirus, 123, 125, 258
Corynosoma magdaleni, 178. See also thorny-headed worms
Coscinocera hercules, 163. See also moths
cospeciation, 13, 16, 17, 42, 215, 263
Costa Rica, 60, 133
counter-current heat exchanger, 195
COVID-19 pandemic, 7, 123, 125, 149, 258, 264
coyotes: diets of, 223, 235–6; evolution of, 222; expanding range and population of, 227, 237; and flukes, 234; human

attempts to eliminate, 224, 227; and itch mites, 225–7; parasite biodiversity, 237; and roundworms, 228–9, 234; sarcoptic mange, 226, 228, 234, 259; social life of, 223; spread of parasites to cities/humans, 221, 238; and tapeworms, 229–33, 234; in urban areas, 228–31, 234–8. *See also* hydatid disease
Coyote/Trickster stories, 222, 223–4, 238
crab lice. *See* lice, pubic
cranial sinuses, 73
Cretaceous era, 193, 248, 263
Cretaceous-Paleocene (K-Pg) boundary, 248, 249
crocodiles, 207, 215
Crohn's disease, 259
cross-immunity, 132
crustacean, 27, 179, 180, 184, 187
Cryptoses, 171. *See also* moths
cystacanth, 179, 180, 182, 184–5, 187

Dama dama, 53. *See also* deer
Darwin, Charles, 84–5, 98, 100, 111, 262
Dawkins, Richard, 255
dazzle effect, 86
deer: black-tailed deer, 52, 132; fallow deer, 53; mule deer, 132; red deer, 53. *See also* white-tailed deer
deer mice, 233
defenses against parasites. *See* host defenses
Delphinidae, 73
Denisovans, 39–40
Dermacentor albipictus, 196. *See also* ticks
dermatitis, 36, 211
Desmana moschata, 61
Desmodus, 119

Desmodus rotundus, 115. *See also* vampire bats
Diaemus youngi, 115. *See also* vampire bats
Dicrocoelium dendriticum, 189. *See also* flatworm flukes
dinosaurs, 239
Diphylla ecaudata, 115. *See also* vampire bats
diphyllobothridians, 142. *See also* tapeworms
Diphyllobothrium, 143–4. *See also* tapeworms
DNA. *See* mitochondrial DNA
dogs (domestic): and ascarid worms, 153; and coyotes, 238; deworming of, 153; and hookworms, 229; and lice, 234–5; and mange, 234–5; and pinworms, 12; and sand fleas, 251; and tapeworms, 230, 231–2. *See also* short-eared (wild) dogs
dolphins, 71, 73, 78, 187, 188
domestic animals. *See* cats (domestic); dogs (domestic)
dominance, 86, 101, 102, 103
Dominican Republic, 246
dopamine, 189
double-edged sword, 109
Dracula (Stoker), 114
drug resistance, 220
ducks: mallards, 180–1, 186; scaup ducks, 181
dung, 167, 171, 172

Echinococcus canadensis, 3, 6, 142, 231–3. *See also* tapeworms
Echinococcus multilocularis, 229–33, 237
Echinococcus spp., 146
echolocation, 73, 119
ecosystems: effect of parasites on, 6–8, 190–1, 260–1, 264,

ecosystems (*continued*) 266; human impact on, 45, 46, 79–80, 138, 219; and pollution, 178; rapid changes to, 126, 203; theories of, 57; and *Trichinella*, 21, 28; urban, *154*. *See also* oceans
ectocommensal, 68
ectoparasites: bat flies as, 246; on beaver beetles, 64; fleas as, 240, 253, 254; and the plague, 34; on vampire bates, 116, 120; on zebras, 88
Ecuador, 205
Edmonton, 227, 230–6
Edmonton Urban Coyote Project, 236
Egyptians, 36
eland, 87
elephants, 101, 105, 164, 167, 205, 217
Elephas maximus, 105
elk: antlers, 100; and brain worms, 133, 137; and giant liver flukes, 52–3; population size, 4, 135. *See also* Irish elk
endoparasites, 250
endosymbionts, 255
Enterobius, 16
Enterobius gregorii, 20
Enterobius vermicularis, 10, 11, 18. *See also* pinworms
environmental change. *See* climate/environmental change
Equus asinus, 83
Equus burchelli, 83
Equus caballus, 83
Equus grevyi, 83
Equus quagga, 83
Equus zebra, 83
Erethizontidae, 19
Eucestoda, 140
Europe: and beaver beetles, 62; consumption, 114; discovery of liver flukes, 52–3, 55; effect of plague, 32–3, 44, 240; evolution of mammals, 19; evolution of primates, 15, 39–40; and giant deer, 97; and infected foxes, 232; parasite surveys in, 231; prehistoric clothing, 38–9; raccoons introduced to, 153; and *Trichinella*, 28
European beavers, 61
evolutionary history. *See under individual parasites and mammals*
extinction: avoided by beetles, 69; caribou, 136, 138; dinosaurs, 259–60; Eurasian beaver, 67; flukes, 54–5; Irish elk, 97–8; moose, 194, 203; muskoxen, 138; Neanderthals, 40; vulnerability to, 193. *See also* mass extinction events
Ezenwa, Vanessa, 104

fallow deer, 53
Fascioloides magna, 46, 52, 108
fear of feces, 158
fevers, 22, 26, 33, 37, 121, 125, 157. *See also* trench fever; yellow fever
fish: and flukes, 206; for human livelihood, 217; as intermediate parasite hosts, 78, 146, 237; and liver flukes, 46; as prey, 73, 145, 188; and tapeworms, 146; and thorny-headed worms, 176; and *Trichinella*, 27
fission-fusion, 223
flatworm flukes, 46, 76–9
flatworms, 46, 206
fleas: anatomy of, 242; as cause of Black Death, 32–4; description of, 240; diets of, 242–3; evolution of, 245–9, 251–3; geographic range of, 253, 265; hosts of, 244–5, 248–54; life cycle of, 244; mating

and reproduction, 243–4; "pre-fleas," 246; sand fleas (Tungids), 249–54; specialized adaptations of, 241–2. See also *Pulex irritans* (human fleas)
flies: bat flies, 246, 247; bot flies, 88, 103; tabanids, 90f, 91–4; true flies, 88; tsetse flies, 89–91
Florida, 115
flukes. *See* blood flukes; flatworm flukes; liver flukes
flying squirrels, 15, 16
foxes, 146, 231–2, 252
Franz Josef Land, 22, 27
frogs, 81, 234
fungi, 58–60, 125, 171, 236, 256, 257

Gammarus lacustris, 179
gangrene, 251
Gasterophilidae, 88
gastropod, 46, 78
gazelles, 100, 109–10
genal comb, 241
gene transcripts, 219
ghost moose, 194–6, 202–3
giant liver flukes. *See* liver flukes
gigantism, 49
giraffes, 200, 207, 217
Glaucomys, 15, 217
Globicephala melas, 72
Glossina spp., 89–91
glucocorticoids, 109
glycocholic acid, 51
Glyptodonts, 119, 253, *254*
goats, 54, 89, 105–6, 132, 133–4, 189
Gondwanaland, 17, 248, 252
gorillas, 15, 32, 34, 41, 42–3
Grant's gazelle, 109
granulomas, 157214
Grassi, Giovanni, 151
great apes. *See under* apes
Great Britain, 97

Greatest Show on Earth, The (Dawkins), 255
greenhouse gases, 263. *See also* climate/environmental change
grizzly bears, 30, 138, 264
grooming: improved methods, 253; and moose, 196, 199–200; to remove beetles, 64; to remove lice, 4, 35; and vampire bats, 120; and zebras, 86, 99. *See also* allogrooming
guar, 100
gulls, 146
Guo, E., 143
gutless wonders, 140

Haldane, J.B.S., 59
Hamilton, Bill, 99, 105, 106, 111
Hamilton-Zuk Hypothesis, 98–9, 105, 106, 111
haplodiploidy, 10–12
hard ticks, 196, 197
harpy eagles, 170
Harrison, Lancelot, 142, 147
Harrison's rule, 139, 142–3, 144, 146, 147
heavy metals, 177
heirloom parasite, 39–40, 42, 254
Hemiceratoides hieroglyphica, 167
hepatic portal system, 208, 212, 214
hepatomegaly, 157
herbivores, 162, 169, 217
hermaphrodites, 51, 140, 206
Hexagonoporus physeteris, 144
Hippocrates, 20, 151
hippos, 217
Holmes, John, 179, 181
Hominidae, 32
Homo habilis, 14
Homo neanderthalensis, 14, 38
Homo sapiens, 14, 32, 38, 39, 40
Hooke, Robert, 240
hookworms, 229, 234

horns: and African buffalo, 104; characteristics of, 100; evolutionary history of, 100–1; and Grant's gazelle, 109–10; and parasites, 105–6; uses of, 101
horse flies, 91–4
host defenses, 110, 184–5, 199–200, 209, 218, 259, 262. *See also* immune responses
host specificity, 13, 18, 132, 215
host switching, 16, 17, 18, 132
house mice, 233
howler monkeys, 43
Hugot, Jean-Pierre, 16
human hunters, 170
human impact on ecosystems. *See under* ecosystems
human infections by parasites: in children, 10, 13; death of child, 157–8; eating polar bear, 22–7; hyatid disease, 230, 233, 259; lice, 36–7, 257–8; roundworms, 228–9; tapeworms, 232; *Toxoplasma*, 189. *See also* Black Death; zoonotic diseases
human pinworms, 13
human-induced climate change. *See* climate/environmental change
Hutchinson, George Evelyn, 57
hydatid cysts, 3, 230, 233
hydatid disease, 230, 233–4, 238, 258, 259, 264
hydatid sand, 230
hyenas, 28, 86, 93
hyperparasites, 121, 124
hypostome, 196
hypothermia, 200
hyraxes, 245

ibexes, 105
immune responses: to acanthocephalans, 179; to blood flukes, 208, 208–9, 214; to brain worms, 132, 134; manipulation of, 209–10, 262; to tapeworms, 3; to thorny-headed worms, 184; to Trichinella, 25. *See also* concomitant immunity; host defenses
immune system, 3, 77, 106, 208, 209, 218–19, 236
importation of exotic animals, 53
India, 15, 19, 20, 105
Indigenous people, 27, 194, 201, 222, 223, 238. *See also* Coyote/Trickster stories
innate immunity, 110
intermediate hosts: altered behaviour of, 184, 260; birds as, 156; cervids as, 231; cockroaches, 182; fish as, 146, 237; of flukes, 234; frogs as, 234; of giant liver flukes, 49, 55; hares as, 235; house mice as, 233; purpose of, 47, 159; rabbits as, 156, 235; rodents as, 155, 156, 158, 258; slugs as, 128; snails as, 128; of tapeworms, 141, 146; and thorny-headed worms, 186; voles as, 235
intestinal infections: by acanthocephalans, 182; by ascarids, 150–2, 156; in beavers, 180; in coyotes, 228–9, 234; in deer, 50–2; in fleas, 32–3; by flukes, 46, 50–2, 210, 212–14, 216, 234; in humans, 23; in raccoons, 156; by roundworms, 23, 228–30, 234; in seals, 178; by tapeworms, 2, 140–1, 144–5, 229, 234; by thorny-headed worms, 176, 178, 180; in whales, 144–5; by whipworms, 28; in wolves, 2
invasive species, 66
Ireland, 97

Irish elk, 96–8, 111
isopods, 142, 184
Ixodes banksi, 68

jaguars, 170, 205, 251
Janzen, Daniel, 65
Japan, 93–4
jays, 200
jiggers, 251
Jolles, Anna, 104
jumping genes, 123

kangaroos, 61
Kenya, 109
kidney damage, 26, 76
Koino, Shimesu, 151–2
K-Pg boundary, 248, 249

lachryphagy, 167
latrines, 155–7, 264
Le Fanu, Joseph Sheridan, 114
Leiodidae, 61
Lemuricola, 16
lemurs, 15, 17, 18
Lepidoptera, 163
Leptinillus, 62
Li, J., 143
lice: and the Black Death, 32–4, 44; body lice, 34, 35f, 36–9; characteristics of, 31; in coyotes, 234; and great apes, 32; head lice, 34–6, 38; and human evolution, 38–40, 43–4; pubic (crab) lice, 34, *35*, 40–3, 258
Linognathus setosus, 234–5
lions, 5, 85, 86, 93, 205
liver flukes: characteristics of, 46; deer as hosts, 50–2, 108; discovery of, 52; as edible "sweetmeats," 46; effects on host mammals, 53–4; geographic location of, 52–3; impacts of habitat change on, 54–6; life cycle of, 46–50; snails as hosts, 46–9, 55–6
Loa loa, 92
Lobocrapsis griseifusa, 168
longhorn steers, 101
long-horned bison, 100
lorises, 15, 17, 18
Loxodonta africana, 105
Luna, 162
lungworms, 106, 137–8, 234
Luzón, 106
Lyme disease, 7, 259, 265
Lymnaea, 47

Macracanthorhynchus hirudinaceus, 179, 186
Macracanthorhynchus ingens, 186
Macropocopris, 61
Macropsyllidae, 249
Madagascar, 15, 17, 167
malaria, 6, 7, 175
mallard ducks, 180–1, 186
mammals: colors of, 81–2; distribution of, 4; grass-eaters and parasites, 189; as parasite hosts, 7–10, 13; and sex chromosomes, 10–12. *See also individual mammals*
Manchurian Candidate, The (Condon), 176
Mandrillus sphinx, 81
mange, 226–7, 234, 259
Manitoba, 201, 232
marine ecosystems. *See* oceans
marine mammals, 78–80, 264. *See also* dolphins; seals; sharks; walruses; whales
marmosets, 15
marsupials, 117, 244, 248, 249
mass extinction events, 193, 263–4
mass stranding events, 72–3, 75, 78–80, 261
Mecoptera, 245

Megaloceros giganteus, 96–7. *See also* Irish elk
Megalongychidae, 168, 170. *See also* sloths
Megatherium, 253. *See also* sloths
melanin, 84, 184
melanization, 184, 185
melanocytes, 84
melons, 76
menigeal worms. *See* brain worms
meninges, 26, 130
meningitis, 26
Mesozoic era, 246, 247, 249
metacercariae, 50
Mexico, 19, 52, 115, 153
mice, 231, 233
Michigan, 45, 51, 52, 202
microbiota, 123–4
Micrographia (Hooke), 240
microtriches, 141
millipedes, 184
mind-control by parasites: ants, 189–90; cockroaches, 182–4; defenses agains, 184–5; and ecosystems, 191; rats, 188–9; scuds, 179–82. *See also* altered behaviour; scuds; thorny-headed worms
Minnesota, 65, 201, 237
miracidia, 47–9, 210
mites, 7, 170, 225–7, 266
mitochondria, 255–6
mitochondrial DNA, 42, 122–3, 248
moles, 248
Moniliformis moniliformis, 182–4, 185
monkeys, 15, 17, 18, 43, 81, 205, 259
monogamous, 216
Montana, 52, 224, 227
moose: and brain worms, 128; ghost moose, 194–6, 202–3; North American stag moose, 97; population increase and decline, 201, 203, 265; size and features of, 194–5; and tapeworms, 4; winter coat of, 195, 200; and winter ticks, 196, 198–203
moose sickness, 128, 133–5, 136, 138
mosquitoes, 5–6
moths: characteristics of, 163; fear of, 162–3; in fur of sloths, 168–72, 171–2; life cycle of, 163; and salt, 164–5, 167–8; tear-feeding species, 167–8; three-toed, 168, 170, 171; two-toed, 168, 170, 171; vampire moths, 164–8; wingspans of, 163
mottephobia, 162–3, 168, 172
mottophilia, 172
mountain beavers, 61
mountain pine beetles, 136, 202, 203
mule deer, 132
Mus musculus, 233
muscle worms, 131, 132, 137
muskoxen, 100, 127, 137–8, 265
muskrats, 61, 180, 181, 186, 260
musth, 105
mutualism/mutualists, 6, 60, 68, 162, 172, 265
Mycobacterium tuberculosis, 114
mycotoxins, 236
myocarditis, 26

nagana, 89–90
nakedness, 38, 43, 260
Nanger granti, 109–10
narwhals, 101
Nasitrema globicephalae, 76–8
natural selection, 147, 187, 257, 262–3
Neanderthals, 38, 39–40
Nematoda, 10, 146

nematodes, 9, 23, 106, 131, 146, 153, 158. *See also* roundworms
neodermis, 48
neosome, 250, 251, 253
Neotropics, 16, 251
Nephridiacanthus longissimus, 186. *See also* thorny-headed worms
neurotransmitter, 184, 187, 189
New Brunswick, 201
New England, 202, 237
New York, 155, 237
New Zealand, 72, 73, 75, 78
Newfoundland, 132, 201
Nicaragua, 59
nits, 34, 36
Noctuidae, 166
North America: and bats, 125; and beaver beetles, 61–2; and brain worms, 133, 137; climate change, 136–7, 203; and coyotes, 221, 222–4, 237; and foxes, 232; and giant liver flukes, 52–5; and horned animals, 100; invasive species, 61–2; and moose, 97, 194; and muscle worms, 132; parasite surveys, 231–2, 236; squirrels and pinworms, 15–16; and *Trichinella*, 28, 29; and white-tailed deer, 55, 131, 135, 137; and wolves, 224
Norway, 22, 67
Nova Scotia, 201
nuptial gift, 166
nurse cells, 25
nutrition, 60, 61, 100, 102, 122
Nycteribiidae, 246, 247. *See also* bat flies

oceans, 71–2, 79–80. *See also* whales
ocelots, 171
octopamine, 184
Odocoileus hemionus, 52

Odocoileus virginianus, 45, 131. *See also* white-tailed deer
Oestridae, 88. *See also* bot flies
omnivores, 29, 150, 153, 235
Ondatra zibethicus, 181
Ontario, 201
oral papilloma virus, 235
orcas, 73, 74, 75
Order Diptera, 88
ornaments: meaning conveyed by, 102–3; and parasites, 95, 103–4, 107–12, 260; and sexual selection, 97–9. *See also* antlers; Darwin, Charles; Hamilton-Zuk Hypothesis; horns; Irish elk; tusks
Oslerus, 234
over dispersed, 218
Ovibos moschatus, 137
ovines, 100
Ovis canadensis, 101
owls, 74, 245
oxpeckers, 200
oxyurids, 10

pacaranas, 19
Pan, 14
Panama, 153, 237
papilloma virus, 235
parasites: compared with predators, 5–6; ecological roles of, 259–66; effects of, 7–8; importance of, 6–8; misconceptions about, 256–7; sexual reproduction of, 261–2; ubiquity of, 58, 162. *See also* acanthocephalans; Ascarids; bat bugs; beaver beetles; beetles; blood flukes; brain worms; commensals; ecosystems: effects of parasites on; flatworm flukes; fleas; flies; hookworms; human infections by parasites; lice;

parasites (*continued*)
liver flukes; moths; muscle worms; mutualism/mutualists; parasitism; pinworms; roundworms; tapeworms; thorny-headed worms; ticks; vampire bats
parasitism: biological prevalence of, 114, 122, 256; and energy/resource acquisition, 58, 256; and formation of life, 255; host-parasite relations, 256–7
Parelaphostrongylus andersoni, 131, 132. *See also* muscle worms
Parelaphostrongylus odocoilei, 132
Parelaphostrongylus tenuis, 128, 132, 133, 135
parthenogenesis, 12, 13–15, 17
Patagonia, 28
pathology, 26, 77, 183
peccaries, 119, 205, 217
pediculosis, 37
Pediculus humanus capitis, 34, 38–42
Pediculus mjobergi, 43
Pediculus schaeffi, 41
Peromyscus maniculatus, 233
P.h. corporis, 34
pheromones, 183, 212
Phoca hispida, 178
photosynthesis, 256
photosynthesizers, 161
Phthirus pubis, 34, 40–2
Phylum Nematoda, 10, 146. *See also* roundworms
Phylum Platyhelminthes, 46, 76, 140, 206. *See also* flatworms; flukes
Physaloptera, 234
Physeter catodon, 144–6
physiology, 4
pigs: and convergent evolution, 189; and fleas, 245, 251; guinea pigs, 53; and ornaments, 101; and roundworms, 150; and thorny-headed worms, 179, 186; and *Trichinella*, 29

pilot whales: characteristics of, 73–4; communication skills, 74–5; infected by parasites, 76–9; mass stranding, 72–3, 75, 78–9

pinworms: diagram of, *11*; evolution of, 14, 17–18, 20; host specificity, 13; in humans, 10, 18, 20; in monkeys, 18; named for, 10; origins of, 17; in primates, 15–17; reproduction and transmission of, 12–13, 15–17; sex determination, 10–12; in squirrels, 15–16

Placentonema gigantissima, 146, 150
plague of 1347. *See* Black Death
plains zebra, 83, 87, 90
Platyhelminthes, 46, 76, 140, 206
Platypsyllus castoris, 62–3, 66–7
platypus, 249
Pleistocene era: giant deer, 97; long-horned bison, 100; muscle worms in caribou, 132; red foxes, 232; squirrel fossils, 18; wolves, 222
pneumonia, 26
polar bears: consumed by Germans, 22; effects of climate change on, 29–30; host to *Trichinella nativa*, 25, 26–7; and *Trichinella nativa*, 23–4
pollution, 71, 75, 79, 178, 263
Polygonoporus giganticus, 144
polygynous, 74, 103, 109
Polymorphus paradoxus, 180, 186
Polyphemus, 162
porcupines, 9, 19–20
praziquantel, 220
predation, 6, 172, 182, 190–1, 206, 256, 260
primates: and beetles, 60; and blood flukes, 207, 217; and lice,

43; and pinworms, 14–20; in South America, 117; species of, 32
proboscis, 176, 179, 182
Procyon lotor, 153, 154. *See also* raccoons
proglottids, 140, 141, 146
pronghorn antelope, 100
pronotal comb, 241
protoscolices, 230–1
protostrongylid roundworms, 137
Protostrongylids, 128, 131
pruritus, 36
pseudocysts, 51, 53, 54
pterosaurs, 247–9
pubic lice, 34, 35, 40–3, 258
puddlers, 167
Pulex irritans, 34, 240, 246
Pyralidae, 171

quaggas, 83
Quebec, 52, 55, 201, 237

rabbits, 4, 53, 156, 223, 230, 234–5, 237, 244
rabies, 67, 114, 125
raccoons, 153–9
rafting, 17, 18, 258
rainforests, 169, 205–6, 263
Rangifer tarandus, 131–2, 136. *See also* caribou
rats: and Black Death, 32–4; and dopamine, 189; and sand fleas, 251; and "suicidal" cockroaches, 182–4, 260; and *Trichinella*, 29
Rattus rattus, 32
ravens, 1, 2, 200
red deer, 53
red-backed voles, 233
redia, 49
reindeer, 103
reptiles, 46, 176, 247
resilin, 241

resource extraction, 55, 127, 136, 263
rhinoceros, 164
ringed seals, 27, 29, 178
rodents: consuming toxins, 221; and coyotes, 223, 230, 237; and fleas, 245, 248, 252, 253; origin of plague, 32; as parasite hosts, 55, 61, 156, 158, 207, 217; populations of, 258; in rainforests, 116; and *Trichinella*, 28; tricked into suicidal acts, 188–9. *See also* beavers; mice; muskrats; porcupines; rats; squirrels
Romans, 36, 150
romantic parasites, 206
roosts/roosting sites, 115, 119–22, 125–6, 264
rotifers, 186–7
roundworms: in coyotes, 228–9; in humans, 10, 150–2, 157–9; in lungs of hosts, 137; in pigs, 150; in raccoons, 155–9; in sloths, 170; in sperm whales, 146. *See also* Ascarids; brain worms; pinworms; *Trichinella nativa*
ruminants, 28, 169
Russia, 22, 61, 164
Russian desman, 61

salt, 164–5, 167–8, 205
San bushmen, 82–3
sand fleas, 250–3
Sanguinicolids, 206
sanguivores, 117
Santa Barbara County, California, 158
sarcoptes scabiei, 225–7
Sarcoptes scabiei, 224, 225, 226
SARS (severe acute respiratory syndrome), 125
Saskatchewan, 153, 201
scabies, 225

scar tissue, 214
scaup ducks, 181
scavengers, 27, 28, 47, 180, 257, 261
scavenging, 26, 29, 256
Schatzgräber weather station, 22
Schistosoma, 207, 216
Schistosoma mansoni, 220
schistosomes, 206–12, 217–20
schistosomiasis, 214, 219
Sciurus, 15
Scotland, 67
scuds (*Gammarus lacustris*), 179–81
Scydosella musawasensis, 59. *See also* beetles
seals, 5, 27, 29, 178
sexual reproduction, 219, 261–2
sexual selection, 97–9, 102, 107, 111, 178, 218. *See also* ornaments
sexy son, 98
sharks, 5, 187, 188
sheep, 54, 56, 89, 100, 101, 133, 134, 189
short-eared (wild) dogs, 170, 205, 217
shrews, 115, 244, 248
Silphopsyllus desmanae, 61
Siphonaptera, 241. *See also* fleas
Sipphopsyllus desmanae, 32. *See also* beetles
Skrjabin, A.S., 144–5
skunks, 153
sloths: and beetles, 60; characteristics of, 168–70; defecation, 170–1; diets of, 169–70, 171–2; and fleas, 253; fur as ecosystem, 171–2; fur color of, 81; moths in fur of, 168–72
slugs, 128, 132, 137
snails: and blood fluke larvae, 211, 217; and flatworm flukes, 189, 234; and liver flukes, 46–9, 55–6; and muscle worms, 128, 132

snow scorpionflies, 245–7
social regurgitation, 124
songs, 221, 243
South America: and fleas, 249, 251, 253; human arrivals in, 43; and pinworms, 18; and schistosomiasis, 216, 217; and sloths, 169
Southeast Asia. *See* Asia
souvenir parasite, 43
sperm whale, 144–6, 147
spermatheca, 244
spider monkeys, 15
Spirorchids, 206
sporocysts, 49, 211
sputum, 151, 152
squirrels: African ground squirrel, 15; flying squirrels, 15–16; fossils of, 18; tree squirrels, 15–16
stag moose, 97
stamina, 4, 107, 183
staphylinids, 60–1. *See also* beetles
stichocytes, 24
Stock's Rule, 147
Stoker, Bram, 114
stomach worms, 234
stranding events. *See* mass stranding events
stress hormones, 106, 109
strobili, 140, 146
Strychnine, 224
sucking lice, 34, *35*, 41, 43
suicide, 73, 179–84, 260
swimmer's itch, 211
sylvatic, 238
symmetry, 98–9, 101, 103, 106
Syncerus caffer, 103–4

tabanids, 90, 91–3
tadpoles, 234
Taenia pisiformis, 234, 237
Taenia saginata, 143, *145*
Taenia serialis, 234

tamarins, 15
Tamiasciurus hudsonicus, 16
Tanzania, 90
tapeworms: broadfish tapeworms, 143–4; in coyotes, 229–33, 234; features of, 140; in foxes, 146, 231–2; lengths of, 143–6; life cycle of, 141–2, 145–6; from raw beef, 143; reproduction of, 140–1, 146; in sloths, 170; in wolves and moose, 2–4
tears as moth food, 167–8
tegument, 48, 141, 177
Tertiary era, 16, 17, 19
testosterone, 109–10
Texas, 101
thermoregulate, 86, 170
thorny-headed worms: ancestors of, 186–7; characteristics of, 176, 177; and environmental pollution, 177–8; evolution of, 177; hosts of, 185–6; life cycle of, 178–9, 185; mating and reproduction, 178. *See also* acanthocephalans; altered behaviour; mind-control by parasites
ticks: characteristics of, 196–7; and climate change, 202–3, 265; life cycle of, 197–8; and moose, 196–203, 265
tiddly-wink flips, 198
Tiputini Biodiversity Station, 205
Titanus giganteus, 59
topis, 87
tourists, 217, 254
Toxacara, 153
Toxascaris leonina, 153, 228, 234, 237
Toxoplasma gondii, 188–9
transmission strategy: of liver flukes, 47–50; manipulation of hosts, 188; of trichinosis, 21
transport hosts, 27, 29
transposons, 122

trappers, 194, 201, 224
tree squirrels, 15–16
Trematoda, 76, 206
trench fever, 6, 37, 121
Trichillum bradyporum, 60
Trichinella, 21, 23–4, 26–30, 257
Trichinella nativa: adaptability and transmission of, 21–2, 28–30; in Arctic, 26–7; behaviour of, 23–5; evolution of, 28; and global warming, 28–30, 263; illness caused by, 22, 25–6
trichinosis, 22–3, 27, 29, 264
Trichophlus spp., 171
Trichuris spp., 28
Trickster. *See* Coyote/Trickster stories
trophic web, 142
true fleas, 62
true flies, 88
Trypanosoma brucei, 89–90
Trypanoxyuris, 17
tsetse flies, 89–91
tuberculosis, 114
Tunga penetrans, 250–3
tungiasis, 250, 254
Tungids, 249–50
turtles, 206, 225
tusks, 101, 105
typhus, 37
Tyrannosaurus rex, 239

Umingmakstrongylus pallikuukensis, 137
Uncinaria stenocephala, 229, 234
United States: and coyotes, 222, 237; infection "hot spots," 52; and invasive species, 66; and pinworms, 20; and raccoons, 153; and vampire bats, 115; and white-tailed deer, 131, 133; and wolves, 224
urbanization, 221, 233, 258
urine, 118, 119

Uroxys gorgon, 60
Ursus maritimus, 22
uterine bell, 178

vaccines, 159, 220
vacuums in nature (*horror vacui*), 161, 162, 173
vagabond's disease, 36
Vámbéry, Ármin, 114
vampire bats: allogrooming, 119, 122, 260; and bat bugs, *120*, 121–2; and blood diet, 117–20, 122, 123–5; evolution of, 115–17; genome of, 123; habitats of, 115; human fear of, 113, 125–6; and prevention of infection, 122–4, 259; regurgitation of food, 120, 124, 260; species of, 115; urination, 118. *See also* bat bugs
vampire legends, 114
vampire moths: characteristics of, 164, 166; evolution of, 166; males as vampires, 164; photos of, *165*; and tear-feeders, 167–8
vascular endothelial growth factor (VGEF), 25
virgin birth, 12
viruses, 91, 122–5, 235, 258, 265
visceral larva migrans, 228, 258
vitamin deficiency, 144
voles, 233
Vulpes lagopus, 232

Wallace, Alfred Russel, 84–5
walruses, 27, 29
wapiti, 52
warthogs, 28
waterholes, 205–6, 217
weapons of competition, 135, 256
weasels, 245
Wellcomia, 19
whales: blue whales, 81; pilot whales, 73–9; sperm whales, 144–6, 147

whipworms, 28
white-footed deer mice, 233
white-tailed deer: and brain worms, 127–30, 133–7, 256; consumed by humans, 45–6; and giant liver flukes, 47, 52–4, 108; and muscle worms, 131, 132; population increase of, 55, 56, 135; in urban areas, 149; and vampire bats, 119
wild asses, 83
wildebeests, 87, 89, 205
wildfires, 136, 202, 203
winter coat, 195, 200
wolves: evolution of, 222; extermination by humans, 224; folk tales about, 224; as predators, 1–2, 135; and tapeworms, 3
World Health Organization, 216

Xenarthra, 168
Xenopsylla cheopis, 32
Xerus inauris, 15

yellow fever, 6, 265. *See also* trench fever
Yersinia pestis, 32–3

zebras, 83. *See also* plains zebra
zebra's stripes: and bot flies, 88–9; as camouflage, 85–6; for communication, 86; evolutionary theories of, 82–5, 93–4, 260; legends of, 82–3; research on, 93–4; and tabanid horse flies, 91–4; to thermoregulate, 86–7; and tsetse flies, 89–91
zombie effect, 24–5, 176
zoonotic diseases, 149, 158–9, 238, 258. *See also* hydatid disease
zoonotic infections, 258
Zuk, Marlene, 99, 105, 106, 111